Grades **3-5**

Math Boosters

Decimals

How to use this book

1. **Let's get started!** Start by writing the date on the top of each page so you can track your progress.

2. **Let's go!** Start with Step 1 and review what you already know.

3. **Don't forget!** Read the "Don't Forget" boxes which contain helpful explanations and examples.

4. **Let's work!** Solve the problems in numerical order step-by-step. Look at the samples problems for help and return to the "Review" box if you need to refresh your knowledge.

5. **Let's check your answers!** After you have finished solving the problems, check your answers and add up your score on each page. If you don't know how to do this, ask your parent or guardian to show you.

6. **Let's get it right!** Once you have finished checking your answers, review any errors to see where you made a mistake and then try again.

To parents

This workbook is designed for children to complete by themselves. By checking their answers and correcting errors on their own, children can strengthen their independence and develop into self-motivated learners.

At Kumon, we believe that each child should do work according to his or her ability, rather than his or her age or grade level. So, if this workbook is too difficult or too easy for your child, please choose another Kumon Math Workbook with an appropriate level of difficulty.

Math Boosters

Grades 3-5 Decimals

● Table of Contents ●

Decimals 1

Date / /

Score /100

Don't Forget!

A way of expressing an amount smaller than 1L in liters.

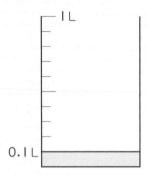

A liter can be equally divided into ten 0.1's of one liter, and this can be read as "zero point one liters."

Numbers like 0.1 or 0.2 are called decimals. "." is called a decimal point.

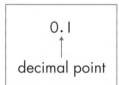

0.1

decimal point

1 Write the answers as decimals to show the amounts. 6 points per question

(1)

They are 2 amounts of 0.1L

[ ] L

(3)

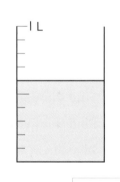

[ ] L

(5)

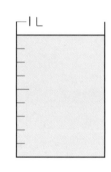

[ ] L

(2)

[ ] L

(4)

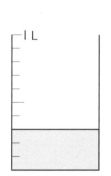

[ ] L

(6)

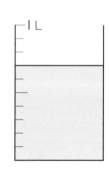

[ ] L

2 Write the answers as decimals to show the amount.

7 points per question

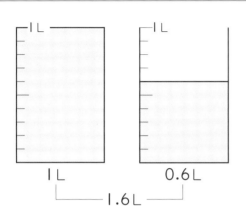

1 L   0.6 L

1.6 L

1.6 L is a combination of 1 L and 0.6 L that is read as "one point six liters."

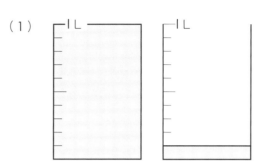

(1)

     L

(2)

     L

(3)

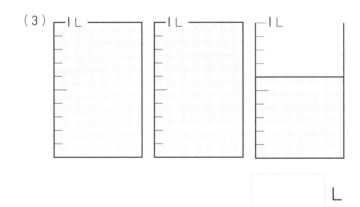

     L

(4)

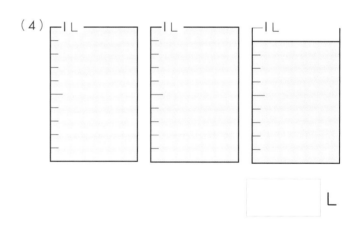

     L

3 1dL = 0.1L. Write the appropriate number in each box.

6 points per question

(1) 5dL = ☐ L

(2) 8dL = ☐ L

(3) 4L 5dL = ☐ L

(4) 7L 2dL = ☐ L

(5) 6L 8dL = ☐ L

(6) 10L 7dL = ☐ L

   5

Decimals 2

Date / /

Score /100

Review STEP **1**   Write the answers as decimals to show the amount.

(1)

☐ L

(2)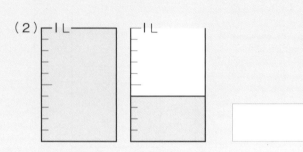

☐ L

1   Write the appropriate number in each box.

5 points per question

Example   How many centimeters long is the tape?

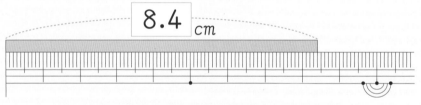

8.4 cm

(1) The length of the tape is 8 cm 4 mm.

(2) 1 cm can be divided equally into ten 1mm.

(3) 1mm is 0.1cm.

(4) The length of the tape is 8.4 cm.

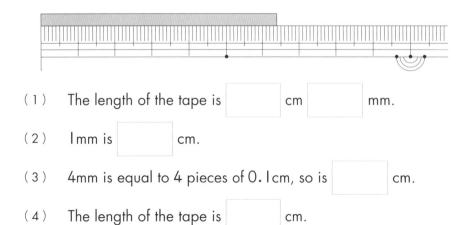

(1) The length of the tape is ☐ cm ☐ mm.

(2) 1mm is ☐ cm.

(3) 4mm is equal to 4 pieces of 0.1cm, so is ☐ cm.

(4) The length of the tape is ☐ cm.

STEP 1-5
Structure of
a Decimal

STEP 6-12
Addition of
Decimals

STEP 13-19
Subtraction of
Decimals

STEP 20-30
Multiplication
of Decimals

STEP 31-40
Decimal ÷
Whole Number

STEP 41-48
Decimal ÷
Decimal

2 **Find the length in centimeters of each point A, B, C, and D as shown from the left end of the ruler.**

5 points per question

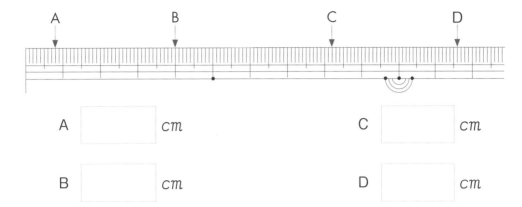

A [ ] cm   C [ ] cm

B [ ] cm   D [ ] cm

3 **Find the length in meters of each point A, B, and C as shown on the tape measure.**

5 points per question

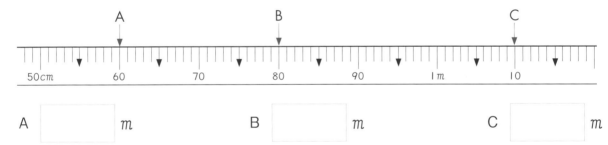

A [ ] m   B [ ] m   C [ ] m

4 **Write the appropriate number in each box.**

5 points per question

(1) 6mm = [ ] cm   (5) 10cm = [ ] m

(2) 10mm = [ ] cm   (6) 1m40cm = [ ] m

(3) 1cm8mm = [ ] cm   (7) 100m = [ ] km

(4) 5cm4mm = [ ] cm   (8) 1km700m = [ ] km

Decimals 3

Review STEP **2**   **How many centimeters long is the tape?**

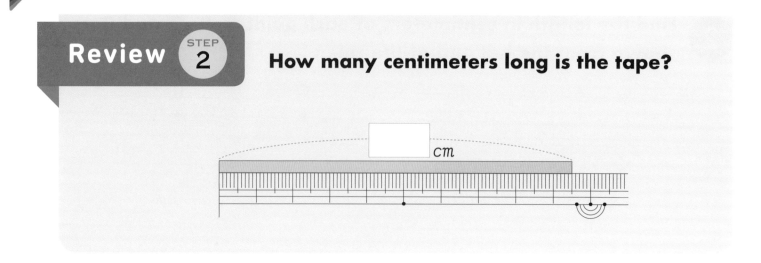

1 Write the numbers based on the points A, B, and C shown on the number lines below.

5 points per question

Example   Numbers are based on the placement of each point on the number line.

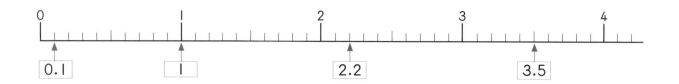

0.1   1   2.2   3.5

(1)

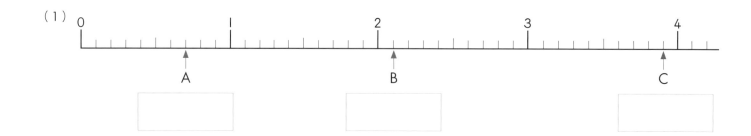

(2)

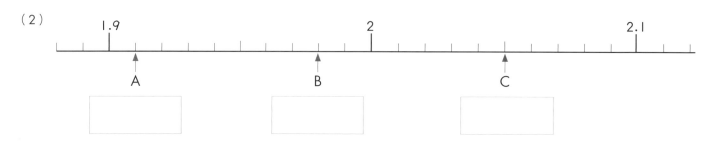

2 Write ↑ to show where the decimals fall on the number line.

5 points per question

0.4	1.3	2.8	3.3

3 Write the larger number by using the number line to compare large and small.

5 points per question

(1) 1.4 and 1.7 ············ ☐   (3) 3.1 and 2.4 ············ ☐

(2) 0.8 and 1.1 ············ ☐

4 Write the numbers from smallest to largest using the number line.

5 points per question

> The difference between large and small numbers can be expressed by using signs like 7 > 6 and 6 < 7. These signs > and < are called inequality signs.

0.12   0   0.08   0.01   ☐ < ☐ < ☐ < ☐

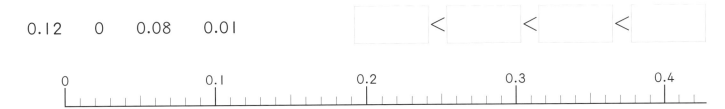

5 Write an inequality sign between the two decimals.   *5 points per question*

(1) 0.4 ☐ 0.6   (2) 7.1 ☐ 6.8   (3) 0.21 ☐ 0.19

Review STEP 3

Write the numbers based on the points A, B, and C shown on the number line below.

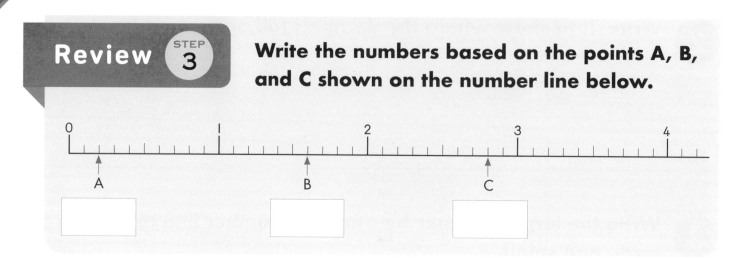

1 Write the number based on the description given.   3 points per question

Example   Answer about 25.67.

- What is the tens place? ——————→ 2
- What is the ones place? ——————→ 5
- What is the tenths place? ——————→ 6
- What is the hundredths place? ———→ 7

> Decimal Place
> In order from the right of decimal point,
> the tenths place
> the hundredths place

(1) The number that is made up of 1 and 0.1.

(2) The number that is made up of 3 and 0.2.

(3) The number that is 0.1 smaller than 1.

(4) The number that is 0.1 larger than 1.

(5) The number that is 0.3 smaller than 2.

(6) The number that is 1.3 larger than 2.

(7) The number of tenths place of 13.24.

(8) The number of hundredths place of 6.78.

STEP 1-5
Structure of
a Decimal

STEP 6-12
Addition of
Decimals

STEP 13-19
Subtraction of
Decimals

STEP 20-30
Multiplication
of Decimals

STEP 31-40
Decimal ÷
Whole Number

STEP 41-48
Decimal ÷
Decimal

2 **How many 0.1 is each number made up of?**

3 points per question

(1) 0.6

(2) 0.9

(3) 1

(4) 1.5

(5) 2

(6) 2.1

(7) 3.3

(8) 11.2

3 **How many 0.01 is each number made up of?**

4 points per question

(1) 0.04

(2) 0.09

(3) 0.12

(4) 0.23

(5) 0.3

(6) 1.25

4 **Write the numbers based on the descriptions.**

4 points per question

(1) The number that is 10 times as 25.67.

(2) The number that is 100 times as 25.67.

(3) The number that is $\frac{1}{10}$ times as 25.67.

(4) The number that is $\frac{1}{100}$ times as 25.67.

(5) The number that is 10 times as 0.6.

(6) The number that is 10 times as 0.04.

(7) The number that is $\frac{1}{10}$ times as 0.6.

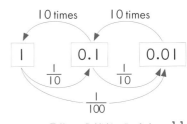

© Kumon Publishing Co., Ltd.   11

Review STEP 4

Write the numbers.

(1) The number makes 1.7 with 1. ·····································

(2) The number that shows how much smaller 1.7 is than 2. ··················

(3) The number that is $\frac{1}{10}$ times as many as 1.7. ·····························

1 **Write the answer in decimals to show the amount.**   12 points

Example

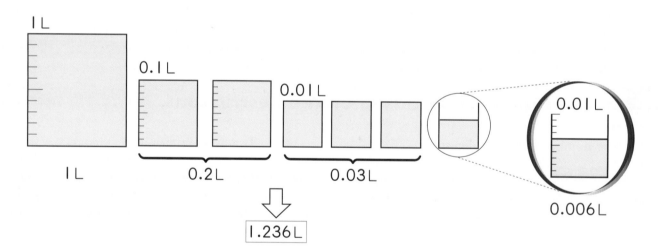

1L   0.2L   0.03L   0.006L

⬇

1.236L

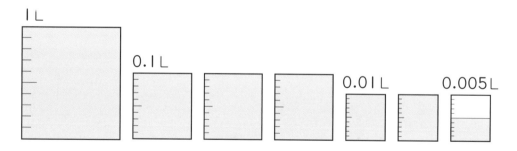

2 Write the numbers.

8 points per question

(1) The number that is made up of four 1s, three 0.1s, seven 0.01s, and five 0.001s.

(2) The number that is made up of two 10s, five 1s, seven 0.1s, four 0.01s, and eight 0.001s.

(3) The number that is made up of three 0.01s, seven 0.001s.

3 Write the numbers based on the points A, B, and C shown on the number line below.

8 points per question

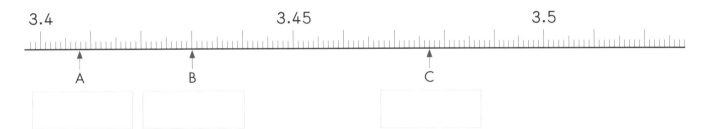

4 Write the numbers.

10 points per question

(1) The number that is 100 times as many as 0.152. ·····················

(2) The number that is 1000 times as many as 0.152. ·····················

(3) The number that is $\frac{1}{100}$ times as many as 5.38. ·····················

(4) The number that is $\frac{1}{1000}$ times as many as 5.38. ·····················

Structure of a Decimal

Date / /

Score /100

Review STEP 1 **How many liters are expressed by the amounts?**

8 points per question

(1)

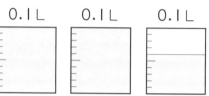
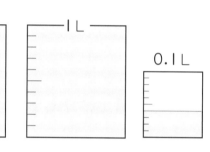

L

(2)

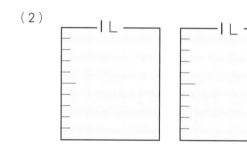

L

Review STEP 2 **How many centimeters long is the tape?** 7 points per question

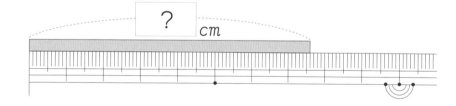

cm

Review STEP 3 **Write the numbers based on the points A, B, and C shown on the number line below.** 7 points per question

A ☐ B ☐ C ☐

Review STEP 4 STEP 5 Write the numbers.

7 points per question

(1) The number that is made up of six 1s, four 0.1s, and five 0.01s.

(2) The number that is made up of ten 10s, seven 1s, three 0.1s, four 0.01s, and two 0.001s.

(3) The number that is made up of fifteen 0.01s.

(4) The number that is 10 times as many as 36.04 .

(5) The number that is 100 times as many as 0.163 .

(6) The number that is $\frac{1}{10}$ times as many as 5.42 .

(7) The number that is $\frac{1}{100}$ times as many as 6.3 .

(8) The number that is $\frac{1}{1000}$ times as many as 5 .

15

Addition of Decimals 1

Date   /   /

Score   /100

Review STEP 5

Write the numbers.

(1) The number that consists of eight 1s and five 0.1s.

(2) The number that consists of three 0.01s and two 0.001s.

1 Add.

5 points per question

Example

$0.3 + 0.4 = 0.7$

0.3 consists of three (3) 0.1s. $\Big)$ $3 + 4 = 7$ Then, you can regard it as seven (7) 0.1s.
0.4 consists of four (4) 0.1s. $0.3 + 0.4 = 0.7$

(1) $0.4 + 0.2 =$

(4) $0.6 + 0.2 =$

(2) $0.7 + 0.1 =$

(5) $0.3 + 0.6 =$

(3) $0.1 + 0.3 =$

(6) $0.2 + 0.3 =$

2 Add.

5 points per question

(1) 1.2 + 0.4 =

(2) 1.3 + 0.5 =

(3) 2.4 + 0.2 =

(4) 3.5 + 0.3 =

(5) 0.6 + 1.2 =

(6) 0.7 + 1.1 =

(7) 0.1 + 2.6 =

(8) 0.2 + 3.7 =

(9) 2.3 + 0.4 =

(10) 4.1 + 0.6 =

(11) 0.3 + 2.5 =

(12) 0.4 + 5.3 =

(13) 0.7 + 3.2 =

(14) 2.6 + 0.2 =

Addition of Decimals 2

Review STEP 6

Add.

(1) 0.5 + 0.2 =

(3) 0.3 + 0.6 =

(2) 2.7 + 0.2 =

(4) 0.4 + 4.2 =

1 Add.

5 points per question

Example 0.7 + 0.3 = 1

$$
\begin{array}{r}
0.7 \\
+0.3 \\
\hline
1.0
\end{array}
$$

Write the answer as 1 and not 1.0.

← Remove 0.

0.5 + 1 = 1.5

$$
\begin{array}{r}
0.5 \\
+1.0 \\
\hline
1.5
\end{array}
$$

← Regard 1.0 as 1.

(1) 0.4 + 0.6 =

(5) 2.7 + 0.3 =

(2) 0.8 + 0.2 =

(6) 1.9 + 0.1 =

(3) 1.6 + 0.4 =

(7) 2.4 + 0.6 =

(4) 0.2 + 1.8 =

2 Add.

5 points per question

(1) $0.3 + 1 =$

(5) $1.4 + 2 =$

(2) $1 + 0.4 =$

(6) $2 + 2.6 =$

(3) $0.6 + 1 =$

(7) $1.3 + 3 =$

(4) $1 + 0.2 =$

(8) $3 + 1.6 =$

3 Add.

5 points per question

(1) $1 + 0.9 =$

(4) $2.5 + 5 =$

(2) $0.8 + 2 =$

(5) $4 + 1.6 =$

(3) $3 + 2.7 =$

Addition of Decimals 3

Review STEP 7

Add.

(1) $0.2 + 0.8 =$

(2) $0.9 + 2.1 =$

(3) $2.4 + 2 =$

(4) $2 + 3.8 =$

---

1 Add.

5 points per question

Example $1.2 + 1.3 = 2.5$

$$\begin{array}{r} 1.2 \\ +1.3 \\ \hline 2.5 \end{array}$$

(1) $1.4 + 1.2 =$

(2) $1.6 + 1.1 =$

(3) $1.3 + 1.5 =$

(4) $1.2 + 1.7 =$

(5) $1.6 + 1.2 =$

(6) $1.5 + 1.2 =$

(7) $1.2 + 1.2 =$

2 Add.

5 points per question

(1) $2.3 + 1.5 =$

(2) $1.3 + 2.6 =$

(3) $2.5 + 1.2 =$

(4) $1.4 + 2.4 =$

(5) $3.2 + 1.7 =$

(6) $1.6 + 3.1 =$

(7) $3.3 + 1.6 =$

(8) $1.5 + 3.4 =$

3 Add.

5 points per question

(1) $2.1 + 3.4 =$

(2) $2.4 + 2.3 =$

(3) $2.6 + 3.1 =$

(4) $4.3 + 3.1 =$

(5) $5.7 + 3.2 =$

Addition of Decimals

Addition of Decimals 4

Date   /   /

Score

/100

Review STEP 8

Add.

(1) $1.2 + 1.5 =$

(2) $2.3 + 1.6 =$

(3) $1.8 + 3.1 =$

(4) $4.2 + 3.6 =$

1 Add.

5 points per question

Example $0.5 + 0.6 = 1.1$

$$
\begin{array}{r}
0.5 \\
+0.6 \\
\hline
1.1
\end{array}
$$

(1) $0.7 + 0.4 =$

(2) $0.8 + 0.5 =$

(3) $0.3 + 0.9 =$

(4) $0.6 + 0.9 =$

(5) $0.8 + 0.7 =$

(6) $0.4 + 0.8 =$

(7) $0.5 + 0.7 =$

2 Add.

5 points per question

(1) $1.6 + 0.5 =$

(2) $0.4 + 1.8 =$

(3) $1.6 + 0.6 =$

(4) $0.3 + 1.9 =$

(5) $1.7 + 0.6 =$

(6) $0.5 + 1.8 =$

(7) $1.9 + 0.3 =$

(8) $0.4 + 1.7 =$

3 Add.

5 points per question

(1) $0.9 + 2.5 =$

(2) $2.4 + 0.8 =$

(3) $2.6 + 0.7 =$

(4) $0.8 + 2.3 =$

(5) $2.2 + 0.9 =$

STEP 10

Addition of Decimals

Addition of Decimals 5

Date   /   /

Score

/100

Review STEP 9

Add.

(1) $0.8 + 0.7 =$ ☐

(2) $1.9 + 0.5 =$ ☐

(3) $1.4 + 0.9 =$ ☐

(4) $0.7 + 0.7 =$ ☐

1 **Add.**

5 points per question

Example $2.4 + 3.7 = 6.1$

$$\begin{array}{r} 2.4 \\ +3.7 \\ \hline 6.1 \end{array}$$

(1) $2.3 + 3.8 =$ ☐

(2) $3.6 + 2.5 =$ ☐

(3) $2.6 + 3.5 =$ ☐

(4) $3.9 + 2.2 =$ ☐

(5) $2.8 + 3.4 =$ ☐

(6) $3.5 + 2.7 =$ ☐

2 Add.

5 points per question

(1) $4.7 + 2.6 =$

(2) $2.5 + 4.8 =$

(3) $4.5 + 2.7 =$

(4) $2.6 + 4.9 =$

(5) $2.6 + 4.6 =$

(6) $4.7 + 2.5 =$

(7) $2.4 + 4.7 =$

(8) $4.4 + 2.8 =$

3 Add.

5 points per question

(1) $5.8 + 2.4 =$

(2) $2.8 + 6.7 =$

(3) $5.4 + 2.7 =$

(4) $2.6 + 6.8 =$

(5) $6.2 + 2.9 =$

(6) $2.5 + 5.6 =$

Addition of Decimals 6

Review STEP 10

Add.

(1) 3.4 + 2.7 = ☐

(3) 4.9 + 2.3 = ☐

(2) 2.6 + 4.6 = ☐

(4) 2.5 + 5.6 = ☐

1 Add.

8 points per question

Example 0.3 + 0.25 = 0.55

```
   0 . 3
 +0 . 2 5
   0 . 5 5
```

1.54 + 12.8 = 14.34

```
   1 . 5 4
 +1 2 . 8
 1 4 . 3 4
```

When you add decimals, align the decimal points.

(1)
```
    0.4
 + 0.3 5
```

(3)
```
    1 5.3
 +  1.2 4
```

(2)
```
    1.6 8
 + 1.3
```

(4)
```
    0.7 3
 + 2 1.2
```

2 Add.

6 points per question

(1) 7 + 3.64 =

(2) 2.27 + 0.9 =

(3) 12.5 + 5.73 =

(4) 4.85 + 21.3 =

(5) 0.5 + 7.84 =

(6) 6.38 + 12.8 =

3 Add.

8 points per question

(1) 2.56 + 2.42 =

(2) 3.64 + 1.35 =

(3) 6.38 + 2.85 =

(4) 4.26 + 1.74 =

Review STEP 11

Add.

(1) $0.3 + 0.65 = $ [   ]

(2) $3.85 + 12.4 = $ [   ]

1 Calculate.

8 points per question

Example

$0.4 + 1.278 = 1.678$

```
  0.4 00
+ 1.278
  1.678
```

$15.63 + 3.865 = 19.495$

```
  15.630
+  3.865
  19.495
```

Don't forget to carry over.

(1)
```
    0.2
+ 0.765
```

(3)
```
    0.04
+ 5.832
```

(2)
```
  1.634
+ 0.24
```

(4)
```
  4.764
+ 2.23
```

2 Calculate.

6 points per question

(1) $0.4 + 0.718 =$ ☐

(4) $2.784 + 15.23 =$ ☐

(2) $6.472 + 2.61 =$ ☐

(5) $3.76 + 4.257 =$ ☐

(3) $4.82 + 1.679 =$ ☐

(6) $5.623 + 4.58 =$ ☐

3 Calculate.

8 points per question

(1) $3.246 + 0.433 =$ ☐

(3) $2.865 + 4.332 =$ ☐

(2) $14.236 + 32.463 =$ ☐

(4) $12.687 + 7.514 =$ ☐

Pay attention to the carry over.

Addition of Decimals

Date / /

Score /100

Review STEP 6 **Add.**

4 points per question

(1) $0.3 + 0.5 =$

(2) $1.2 + 0.4 =$

(3) $0.7 + 4.2 =$

(4) $0.6 + 0.2 =$

Review STEP 7 **Add.**

4 points per question

(1) $1.4 + 0.6 =$

(2) $1.7 + 3 =$

(3) $2 + 2.4 =$

(4) $0.2 + 2.8 =$

(5) $0.8 + 4 =$

(6) $0.8 + 1.2 =$

Review STEP 8 **Add.**

4 points per question

(1) $1.2 + 1.6 =$

(2) $2.4 + 1.5 =$

(3) $1.8 + 4.1 =$

(4) $2.6 + 3.2 =$

Review STEP **9** **Add.**

4 points per question

(1) $0.5 + 0.6 =$

(3) $0.7 + 1.5 =$

(2) $1.7 + 0.4 =$

(4) $1.8 + 0.4 =$

Review STEP **10** **Add.**

4 points per question

(1) $3.5 + 2.6 =$

(3) $4.8 + 2.3 =$

(2) $2.7 + 3.4 =$

(4) $2.5 + 5.7 =$

Review STEP **11** STEP **12** **Add.**

3 points per question

(1)
$$\begin{array}{r} 0.4 \\ + 0.35 \\ \hline \end{array}$$

(3)
$$\begin{array}{r} 3.847 \\ + 1.754 \\ \hline \end{array}$$

(2)
$$\begin{array}{r} 3.75 \\ + 12.6 \\ \hline \end{array}$$

(4)
$$\begin{array}{r} 15.632 \\ + 8.376 \\ \hline \end{array}$$

Subtraction of Decimals 1

Date   /   /

Score   /100

Review STEP 6

Add.

(1) $1.2 + 0.3 =$ ⬚

(3) $0.6 + 1.2 =$ ⬚

(2) $3.5 + 0.2 =$ ⬚

(4) $0.7 + 1.2 =$ ⬚

1 Subtract.

5 points per question

Example $0.8 - 0.3 = 0.5$

0.8 is made up of eight (8) 0.1s.
0.3 is made up of three (3) 0.1s.
$8 - 3 = 5$
Then, 0.5 can be seen as
five (5) 0.1s.

$1.5 - 0.3 = 1.2$

1.5 is made up of fifteen (15) 0.1s.
0.3 is made up of three (3) 0.1s.
$15 - 3 = 12$
Then, 1.2 can be seen as
twelve (12) 0.1s.

(1) $0.5 - 0.3 =$ ⬚

(4) $0.6 - 0.3 =$ ⬚

(2) $0.4 - 0.1 =$ ⬚

(5) $0.9 - 0.7 =$ ⬚

(3) $0.8 - 0.2 =$ ⬚

(6) $0.2 - 0.1 =$ ⬚

2 Subtract.

5 points per question

(1) $1.5 - 0.4 =$

(2) $1.4 - 0.1 =$

(3) $1.8 - 0.6 =$

(4) $2.7 - 0.6 =$

(5) $2.5 - 0.3 =$

(6) $2.3 - 0.2 =$

(7) $2.6 - 0.3 =$

(8) $2.6 - 0.5 =$

(9) $3.7 - 0.4 =$

(10) $3.6 - 0.3 =$

(11) $4.6 - 0.1 =$

(12) $4.8 - 0.7 =$

(13) $5.4 - 0.3 =$

(14) $5.8 - 0.5 =$

Subtraction of Decimals

Subtraction of Decimals 2

Review STEP 13

Subtract.

(1) $0.8 - 0.6 =$

(3) $1.6 - 0.4 =$

(2) $0.7 - 0.2 =$

(4) $3.8 - 0.7 =$

1 Subtract.

5 points per question

Example

$1.2 - 0.2 = 1$

$$\begin{array}{r} 1.2 \\ -0.2 \\ \hline 1.0 \end{array}$$

$1.4 - 1 = 0.4$

$$\begin{array}{r} 1.4 \\ +1.0 \\ \hline 0.4 \end{array}$$

(1) $1.3 - 0.3 =$

(4) $7.3 - 4.3 =$

(2) $2.7 - 1.7 =$

(5) $5.4 - 3.4 =$

(3) $4.6 - 2.6 =$

(6) $8.5 - 6.5 =$

2 Subtract.

5 points per question

(1) $1.3 - 1 =$

(2) $1.6 - 1 =$

(3) $2.5 - 2 =$

(4) $3.4 - 3 =$

(5) $3.8 - 2 =$

(6) $3.2 - 2 =$

(7) $5.7 - 3 =$

(8) $6.7 - 4 =$

(9) $4.6 - 2 =$

(10) $5.8 - 3 =$

(11) $7.8 - 4 =$

(12) $7.6 - 3 =$

(13) $6.7 - 1 =$

(14) $8.9 - 2 =$

Subtraction of Decimals 3

Review STEP 14

Subtract.

(1) $2.6 - 0.6 =$

(3) $6.4 - 2 =$

(2) $4.3 - 2.3 =$

(4) $3.5 - 3 =$

1 Subtract.

5 points per question

Example $1.2 - 0.5 = 0.7$

$$
\begin{array}{r}
1.2 \\
-0.5 \\
\hline
0.7
\end{array}
$$

(1) $1.3 - 0.8 =$

(4) $1.7 - 0.9 =$

(2) $1.4 - 0.6 =$

(5) $1.5 - 0.7 =$

(3) $1.1 - 0.3 =$

(6) $1 - 0.5 =$

2 Subtract.

5 points per question

(1) $2.3 - 0.5 =$

(2) $2.2 - 0.6 =$

(3) $2.5 - 0.8 =$

(4) $2 - 0.9 =$

(5) $3.4 - 0.7 =$

(6) $3.2 - 0.6 =$

(7) $3.2 - 0.3 =$

(8) $3 - 0.7 =$

(9) $4.1 - 0.6 =$

(10) $4.8 - 0.9 =$

(11) $4.7 - 0.8 =$

(12) $4 - 0.5 =$

(13) $5.2 - 0.9 =$

(14) $5.5 - 0.7 =$

Review STEP 15

Subtract.

(1) $1.2 - 0.4 =$ ☐

(3) $3.5 - 0.6 =$ ☐

(2) $2.4 - 0.7 =$ ☐

(4) $5.7 - 0.9 =$ ☐

1 Subtract.

5 points per question

Example $1.6 - 1.2 = 0.4$

$$\begin{array}{r} 1.6 \\ -1.2 \\ \hline 0.4 \end{array}$$

(1) $1.6 - 1.4 =$ ☐

(4) $1.4 - 1.3 =$ ☐

(2) $1.5 - 1.4 =$ ☐

(5) $1.9 - 1.3 =$ ☐

(3) $1.8 - 1.2 =$ ☐

(6) $1.7 - 1.1 =$ ☐

2 Subtract.

5 points per question

(1) $2.9 - 1.5 =$

(2) $2.8 - 2.3 =$

(3) $3.9 - 2.5 =$

(4) $3.8 - 3.7 =$

(5) $4.7 - 2.3 =$

(6) $4.6 - 4.4 =$

(7) $5.7 - 5.4 =$

(8) $6.8 - 4.6 =$

(9) $6.9 - 5.1 =$

(10) $7.9 - 2.7 =$

3 Subtract.

5 points per question

(1) $16.5 - 4.3 =$

(2) $15.8 - 3.8 =$

(3) $15.9 - 11.7 =$

(4) $14.7 - 8.1 =$

Review STEP 16

Subtract.

(1) $1.8 - 1.4 =$ ☐

(3) $6.7 - 5.2 =$ ☐

(2) $5.6 - 4.3 =$ ☐

---

1 **Subtract.**

5 points per question

Example $5.2 - 3.7 = 1.5$

$$\begin{array}{r} 5.2 \\ -3.7 \\ \hline 1.5 \end{array}$$

(1) $5.7 - 1.9 =$ ☐

(4) $6.7 - 2.9 =$ ☐

(2) $5.7 - 2.9 =$ ☐

(5) $6.7 - 3.9 =$ ☐

(3) $5.7 - 3.8 =$ ☐

(6) $6.7 - 4.8 =$ ☐

STEP 1-5
Structure of
a Decimal

STEP 6-12
Addition of
Decimals

STEP 13-19
Subtraction of
Decimals

STEP 20-30
Multiplication
of Decimals

STEP 31-40
Decimal ÷
Whole Number

STEP 41-48
Decimal ÷
Decimal

2 **Subtract.**

5 points per question

(1) $5.2 - 3.9 =$

(2) $5.2 - 2.9 =$

(3) $6.2 - 4.8 =$

(4) $6.2 - 3.8 =$

(5) $4.1 - 3.6 =$

(6) $4.1 - 3.7 =$

(7) $3.7 - 2.9 =$

(8) $3.2 - 2.9 =$

(9) $3.2 - 2.5 =$

(10) $2.1 - 1.9 =$

3 **Subtract.**

5 points per question

(1) $3 - 1.7 =$

(2) $4 - 1.2 =$

(3) $3 - 2.5 =$

(4) $5 - 4.6 =$

41

Subtraction of Decimals 6

Review STEP 17

Subtract.

(1) $5.7 - 1.8 =$ ☐

(2) $6.7 - 3.8 =$ ☐

1 Subtract.

6 points per question

Example

$$7.85 - 2.64 = 5.21$$

$$
\begin{array}{r}
7.85 \\
-2.64 \\
\hline
5.21
\end{array}
$$

$$2.3 - 1.16 = 1.14$$

$$
\begin{array}{r}
2.3 \\
-1.16 \\
\hline
1.14
\end{array}
$$

When you subtract decimals, align the decimal points.

(1)
$$
\begin{array}{r}
7.84 \\
-3.52
\end{array}
$$

(3)
$$
\begin{array}{r}
4.56 \\
-2.31
\end{array}
$$

(2)
$$
\begin{array}{r}
8.46 \\
-2.3
\end{array}
$$

(4)
$$
\begin{array}{r}
0.17 \\
-0.04
\end{array}
$$

2 Subtract.

6 points per question

(1) $5.82 - 2.58 =$

(2) $8.46 - 3.75 =$

(3) $8.46 - 2.5 =$

(4) $2.38 - 1.6 =$

(5) $6.24 - 4.38 =$

(6) $7.46 - 0.57 =$

(7) $5.21 - 4.9 =$

(8) $3.85 - 0.9 =$

3 Subtract.

7 points per question

(1) $3.7 - 1.24 =$

(2) $3.2 - 0.28 =$

(3) $2 - 0.06 =$

(4) $3 - 0.42 =$

Subtraction of Decimals 7

Review STEP 18

Subtract.

(1) 7.64 − 3.52 = ☐

(2) 8.46 − 2.4 = ☐

1 Subtract.

6 points per question

Example

$$4.567 − 1.232 = 3.335$$

$$\begin{array}{r} 4.567 \\ -1.232 \\ \hline 3.335 \end{array}$$

$$2.3 − 1.166 = 1.134$$

$$\begin{array}{r} 2.300 \\ -1.166 \\ \hline 1.134 \end{array}$$

(1) $\begin{array}{r} 7.856 \\ -5.632 \\ \hline \end{array}$

(3) $\begin{array}{r} 7.865 \\ -5.75 \\ \hline \end{array}$

(2) $\begin{array}{r} 6.584 \\ -5.463 \\ \hline \end{array}$

(4) $\begin{array}{r} 8.437 \\ -8.4 \\ \hline \end{array}$

2 Subtract.

6 points per question

(1) $6.234 - 4.872 =$

(5) $3.582 - 2.6 =$

(2) $7.652 - 5.761 =$

(6) $5.738 - 5.7 =$

(3) $6.234 - 4.75 =$

(7) $3.582 - 2.691 =$

(4) $7.652 - 6.78 =$

(8) $5.738 - 0.059 =$

3 Subtract.

7 points per question

(1) $2.4 - 1.276 =$

(3) $8.76 - 7.879 =$

(2) $2.48 - 1.495 =$

(4) $3 - 2.764 =$

Subtraction of Decimals

Review STEP **13**   **Subtract.**

4 points per question

(1) $0.4 - 0.2 = $

(3) $5.4 - 0.3 = $

(2) $0.8 - 0.5 = $

(4) $4.7 - 0.5 = $

Review STEP **14**   **Subtract.**

4 points per question

(1) $1.4 - 0.4 = $

(4) $6.7 - 3 = $

(2) $7.4 - 5.4 = $

(5) $7.8 - 7 = $

(3) $2.3 - 2 = $

Review STEP **15**   **Subtract.**

5 points per question

(1) $1.2 - 0.8 = $

(3) $5.8 - 0.9 = $

(2) $1 - 0.7 = $

(4) $7.2 - 0.9 = $

Review STEP 16 STEP 17 — Subtract.

5 points per question

(1) $1.7 - 1.3 =$ ⬚

(3) $6.7 - 4.9 =$ ⬚

(2) $8 - 6.3 =$ ⬚

(4) $3.4 - 2.7 =$ ⬚

Review STEP 18 — Subtract.

4 points per question

(1)
```
  7.6 5
- 3.4 2
```

(3)
```
  3.4
- 1.2 6
```

(2)
```
  6.2 3
- 2.5
```

Review STEP 19 — Subtract.

4 points per question

(1)
```
  7.8 6
- 5.9 5 3
```

(3)
```
  4.3
- 3.5 4 7
```

(2)
```
  6.2 5 8
- 4.6 8
```

STEP **20**

Multiplication of Decimals

Mental Calculation 1:
Decimal × Whole Number

Date   /   /

Score   /100

Review STEP 6

Add.

(1) $0.7 + 3.2 =$

(3) $4.2 + 0.5 =$

(2) $2.3 + 0.4 =$

(4) $2.6 + 0.2 =$

1 Multiply.

5 points per question

Example $0.2 × 4 = 0.8$

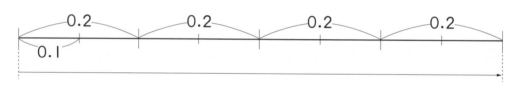

0.2 ·············· two 0.1 s

$0.2 × 4$ ······ $(2 × 4)$ 0.1 s

(1) $0.4 × 2 =$

(4) $0.8 × 6 =$

(2) $0.4 × 8 =$

(5) $0.5 × 4 =$

(3) $0.6 × 4 =$

Write 2 and not 2.0.

STEP 1-5
Structure of
a Decimal

STEP 6-12
Addition of
Decimals

STEP 13-19
Subtraction of
Decimals

STEP 20-30
Multiplication
of Decimals

STEP 31-40
Decimal ÷
Whole Number

STEP 41-48
Decimal ÷
Decimal

2 Multiply.

5 points per question

(1) $1.2 \times 2 =$

(2) $1.2 \times 3 =$

(3) $1.2 \times 4 =$

(4) $1.2 \times 5 =$

(5) $1.4 \times 2 =$

(6) $1.4 \times 3 =$

(7) $2.3 \times 2 =$

(8) $2.3 \times 3 =$

(9) $3.2 \times 2 =$

(10) $3.2 \times 3 =$

3 Multiply.

5 points per question

(1) $3.1 \times 4 =$

(2) $2.1 \times 6 =$

(3) $3.2 \times 4 =$

(4) $2.1 \times 5 =$

(5) $6.4 \times 2 =$

Review STEP 20

Multiply.

(1) $0.8 \times 7 =$ ☐

(3) $3.1 \times 5 =$ ☐

(2) $2.3 \times 3 =$ ☐

(4) $3.1 \times 8 =$ ☐

1 Multiply.

5 points per question

Example

$$0.4 \times 20 = 8$$

⇩

0.4 ·············· four 0.1 s

0.4×20 ············ (4×20) 0.1 s

$$0.4 \times 200 = 80$$

⇩

0.4 ·············· four 0.1 s

0.4×200 ········ (4×200) 0.1 s

(1) $0.3 \times 30 =$ ☐

(4) $0.2 \times 300 =$ ☐

(2) $0.3 \times 200 =$ ☐

(5) $0.5 \times 20 =$ ☐

(3) $0.2 \times 40 =$ ☐

2 Multiply.

5 points per question

(1) 0.5 × 40 =

(2) 0.6 × 40 =

(3) 1.2 × 20 =

(4) 1.2 × 40 =

(5) 2.3 × 20 =

(6) 2.3 × 30 =

(7) 3.6 × 10 =

(8) 3.6 × 100 =

(9) 2.4 × 10 =

(10) 2.4 × 100 =

3 Multiply.

5 points per question

(1) 0.6 × 400 =

(2) 0.6 × 800 =

(3) 1.2 × 200 =

(4) 2.3 × 300 =

(5) 3.6 × 1000 =

STEP **22**

Multiplication of Decimals

Decimal ×
Whole Number 1

Date / /

Score /100

Review STEP 21

Multiply.

(1) $0.3 \times 40 =$ ☐

(3) $1.5 \times 200 =$ ☐

(2) $1.6 \times 50 =$ ☐

(4) $3.5 \times 400 =$ ☐

(1)

1 **Calculate.**

6 points per question

Example

$$\begin{array}{r} 1.4 \\ \times \quad 3 \\ \hline 4.2 \end{array}$$

(1) First, calculate 14×3. ⟶

$$\begin{array}{r} 1\ 4 \\ \times \quad 3 \\ \hline 4\ 2 \end{array}$$

(2) Then, place the decimal point. ⟶

$$\begin{array}{r} 1.4 \\ \times \quad 3 \\ \hline 4.2 \end{array}$$

(1)
$$\begin{array}{r} 0.7 \\ \times \quad 4 \\ \hline 2.8 \end{array}$$

(3)
$$\begin{array}{r} 1.2 \\ \times \quad 6 \\ \hline \end{array}$$

(5)
$$\begin{array}{r} 2.4 \\ \times \quad 3 \\ \hline \end{array}$$

(2)
$$\begin{array}{r} 0.9 \\ \times \quad 7 \\ \hline \square.\square \end{array}$$

(4)
$$\begin{array}{r} 1.3 \\ \times \quad 4 \\ \hline \end{array}$$

2 Calculate.

5 points per question

(1)
$$\begin{array}{r} 1.3 \\ \times\ \ 8 \\ \hline \end{array}$$

(4)
$$\begin{array}{r} 5.4 \\ \times\ \ 3 \\ \hline \end{array}$$

(7)
$$\begin{array}{r} 12.4 \\ \times\ \ \ \ 2 \\ \hline \end{array}$$

(2)
$$\begin{array}{r} 2.9 \\ \times\ \ 6 \\ \hline \end{array}$$

(5)
$$\begin{array}{r} 3.5 \\ \times\ \ 4 \\ \hline .0 \end{array}$$

Remove 0!

(8)
$$\begin{array}{r} 21.4 \\ \times\ \ \ \ 3 \\ \hline \end{array}$$

(3)
$$\begin{array}{r} 4.2 \\ \times\ \ 7 \\ \hline \end{array}$$

(6)
$$\begin{array}{r} 5.8 \\ \times\ \ 6 \\ \hline \end{array}$$

3 Calculate.

6 points per question

(1)
$$\begin{array}{r} 2.7 \\ \times\ \ 8 \\ \hline \end{array}$$

(3)
$$\begin{array}{r} 2.8 \\ \times\ \ 9 \\ \hline \end{array}$$

(5)
$$\begin{array}{r} 42.5 \\ \times\ \ \ \ 6 \\ \hline \end{array}$$

Remove 0!

(2)
$$\begin{array}{r} 3.9 \\ \times\ \ 8 \\ \hline \end{array}$$

(4)
$$\begin{array}{r} 30.7 \\ \times\ \ \ \ 8 \\ \hline \end{array}$$

STEP 23

Multiplication of Decimals

Decimal ×
Whole Number 2

Date   /   /

Score   /100

Review STEP 22

Calculate.

(1)
```
    4.3
 ×    5
```

(2)
```
    2.8
 ×    8
```

(3)
```
   24.6
 ×     4
```

1 Calculate.

6 points per question

Example

```
   1.16
 ×    3
   3.48
```

(1) First, calculate 116×3. ⟶
```
   116
 ×   3
   348
```

(2) Then, place the decimal point. ⟶
```
   1.16
 ×    3
   3.48
```

(1)
```
   1.23
 ×    2
  □.□□
```

(3)
```
   2.14
 ×    4
```

(5)
```
   1.08
 ×    4
```

(2)
```
   1.24
 ×    3
```

(4)
```
   3.26
 ×    3
```

2 Calculate.

5 points per question

(1)
$$\begin{array}{r} 0.75 \\ \times\ \ \ \ 3 \\ \hline \end{array}$$

(4)
$$\begin{array}{r} 4.23 \\ \times\ \ \ \ 4 \\ \hline \end{array}$$

(7)
$$\begin{array}{r} 1.28 \\ \times\ \ \ \ 4 \\ \hline \end{array}$$

(2)
$$\begin{array}{r} 0.64 \\ \times\ \ \ \ 6 \\ \hline \end{array}$$

(5)
$$\begin{array}{r} 3.06 \\ \times\ \ \ \ 7 \\ \hline \end{array}$$

(8)
$$\begin{array}{r} 1.28 \\ \times\ \ \ \ 5 \\ \hline 0 \end{array}$$

↑
Remove 0!

(3)
$$\begin{array}{r} 2.76 \\ \times\ \ \ \ 3 \\ \hline \end{array}$$

(6)
$$\begin{array}{r} 4.03 \\ \times\ \ \ \ 5 \\ \hline \end{array}$$

3 Calculate.

6 points per question

(1)
$$\begin{array}{r} 3.26 \\ \times\ \ \ \ 5 \\ \hline \end{array}$$

(3)
$$\begin{array}{r} 2.74 \\ \times\ \ \ \ 9 \\ \hline \end{array}$$

(5)
$$\begin{array}{r} 4.56 \\ \times\ \ \ \ 7 \\ \hline \end{array}$$

(2)
$$\begin{array}{r} 3.82 \\ \times\ \ \ \ 5 \\ \hline \end{array}$$

(4)
$$\begin{array}{r} 3.84 \\ \times\ \ \ \ 6 \\ \hline \end{array}$$

STEP **24**

Multiplication of Decimals
Decimal ×
Whole Number 3

Date   /   /

Score

/100

Review STEP **23**

Calculate.

(1)
```
   1.28
 ×    3
```

(2)
```
   2.14
 ×    2
```

(3)
```
   3.82
 ×    6
```

1 Calculate.

6 points per question

Example

```
   1.36
 ×   12
   272
  136
 16.32
```

First, calculate $136 × 12$.
Then, place the decimal point.

(1)
```
   1.36
 ×   13
```

(3)
```
   1.48
 ×   13
```

(5)
```
   2.13
 ×   17
```

(2)
```
   1.42
 ×   14
```

(4)
```
   1.56
 ×   16
```

2 Calculate.

5 points per question

(1)
$$2.13 \times 25$$

(4)
$$2.23 \times 28$$

(7)
$$3.24 \times 34$$

(2)
$$1.56 \times 24$$

(5)
$$3.15 \times 26$$

(8)
$$5.03 \times 38$$

(3)
$$3.06 \times 27$$

(6)
$$1.47 \times 32$$

3 Calculate.

6 points per question

(1)
$$1.72 \times 54$$

(3)
$$2.21 \times 67$$

(5)
$$4.23 \times 78$$

(2)
$$1.24 \times 46$$

(4)
$$3.47 \times 84$$

Multiplication of Decimals
Decimal ×
Whole Number 4

Review STEP 24

Calculate.

(1)
```
    1.36
×     12
```

(2)
```
    3.23
×     25
```

(3)
```
    4.23
×     24
```

1 Calculate.

5 points per question

Example

```
  0.36        ×100          36
×   12      ─────────→    ×  12
    72                       72
  36                       36
  4.32      ←─────────     432
              ÷100
```

(1)
```
    0.36
×     14
```

(3)
```
    0.32
×     16
```

(5)
```
    0.66
×     14
```

(2)
```
    0.28
×     13
```

(4)
```
    0.34
×     12
```

2 Calculate.

5 points per question

(1)
$$\begin{array}{r} 0.63 \\ \times\ 16 \\ \hline \end{array}$$

(2)
$$\begin{array}{r} 0.39 \\ \times\ 32 \\ \hline \end{array}$$

(3)
$$\begin{array}{r} 0.46 \\ \times\ 28 \\ \hline \end{array}$$

(4)
$$\begin{array}{r} 0.24 \\ \times\ 47 \\ \hline \end{array}$$

(5)
$$\begin{array}{r} 0.48 \\ \times\ 46 \\ \hline \end{array}$$

(6)
$$\begin{array}{r} 0.52 \\ \times\ 26 \\ \hline \end{array}$$

(7)
$$\begin{array}{r} 0.32 \\ \times\ 46 \\ \hline \end{array}$$

(8)
$$\begin{array}{r} 0.73 \\ \times\ 25 \\ \hline \end{array}$$

(9)
$$\begin{array}{r} 0.63 \\ \times\ 32 \\ \hline \end{array}$$

(10)
$$\begin{array}{r} 0.35 \\ \times\ 33 \\ \hline \end{array}$$

(11)
$$\begin{array}{r} 0.57 \\ \times\ 35 \\ \hline \end{array}$$

(12)
$$\begin{array}{r} 0.42 \\ \times\ 53 \\ \hline \end{array}$$

(13)
$$\begin{array}{r} 0.44 \\ \times\ 31 \\ \hline \end{array}$$

(14)
$$\begin{array}{r} 0.64 \\ \times\ 45 \\ \hline \end{array}$$

(15)
$$\begin{array}{r} 0.84 \\ \times\ 72 \\ \hline \end{array}$$

Multiplication of Decimals
Whole Number ×
Decimal 1

Date　　/　　/

Score　　　　/100

Review STEP 25

Calculate.

(1)　　0.36
　　×　12

(2)　　0.48
　　×　34

(3)　　0.62
　　×　47

1 Calculate.

6 points per question

Example

　　16
　×0.3
　　4.8

(1) First, calculate 16 × 3. ⟶

　　16
　×　3
　　48

(2) Then, place the decimal point. ⟶

　　16
　×0.3
　　4.8

(1)　　12
　×　0.6

(3)　　12
　×　0.8

(5)　　18
　×　0.3

(2)　　12
　×　0.7

(4)　　16
　×　0.4

2 Calculate.

5 points per question

(1)
$$\begin{array}{r} 2\ 5 \\ \times\ 0.7 \\ \hline \end{array}$$

(2)
$$\begin{array}{r} 3\ 4 \\ \times\ 0.4 \\ \hline \end{array}$$

(3)
$$\begin{array}{r} 3\ 6 \\ \times\ 0.4 \\ \hline \end{array}$$

(4)
$$\begin{array}{r} 3\ 7 \\ \times\ 0.4 \\ \hline \end{array}$$

(5)
$$\begin{array}{r} 4\ 2 \\ \times\ 0.7 \\ \hline \end{array}$$

(6)
$$\begin{array}{r} 4\ 3 \\ \times\ 0.5 \\ \hline \end{array}$$

(7)
$$\begin{array}{r} 5\ 4 \\ \times\ 0.3 \\ \hline \end{array}$$

(8)
$$\begin{array}{r} 5\ 6 \\ \times\ 0.3 \\ \hline \end{array}$$

3 Calculate.

6 points per question

(1)
$$\begin{array}{r} 1\ 8 \\ \times\ 0.7 \\ \hline \end{array}$$

(2)
$$\begin{array}{r} 2\ 5 \\ \times\ 0.4 \\ \hline \end{array}$$

Remove 0!

(3)
$$\begin{array}{r} 2\ 5 \\ \times\ 0.9 \\ \hline \end{array}$$

(4)
$$\begin{array}{r} 2\ 9 \\ \times\ 0.4 \\ \hline \end{array}$$

(5)
$$\begin{array}{r} 2\ 6 \\ \times\ 0.8 \\ \hline \end{array}$$

Multiplication of Decimals
Whole Number ×
Decimal 2

Date   /   /

Score

/100

Review STEP 26

Calculate.

(1)
```
    1 2
×   0.6
```

(2)
```
    1 8
×   0.7
```

(3)
```
    4 3
×   0.4
```

1 **Calculate.**

6 points per question

Example

```
    1 6
×   1.4
─────────
    6 4
  1 6
─────────
  2 2.4
```

First, calculate 16 × 14.

Then, place the decimal point.

(1)
```
    1 6
×   1.2
```
```
┌──┬──┐
└──┴──┘
┌──┬──┐
└──┴──┘
┌──┬──┬──┐
└──┴──┴──┘
```

(3)
```
    2 6
×   1.3
```

(5)
```
    3 4
×   1.6
```

(2)
```
    1 7
×   1.3
```

(4)
```
    2 3
×   1.5
```

2 Calculate.

5 points per question

(1)
```
    2 3
×   2.1
```

(2)
```
    2 4
×   2.8
```

(3)
```
    2 9
×   2.7
```

(4)
```
    2 6
×   3.2
```

(5)
```
    3 6
×   3.6
```

(6)
```
    4 8
×   3.9
```

(7)
```
    3 4
×   4.3
```

(8)
```
    3 7
×   4.5
```

3 Calculate.

6 points per question

(1)
```
    3 8
×   5.7
```

(2)
```
    4 6
×   6.5
```

(3)
```
    5 2
×   7.8
```

(4)
```
    5 9
×   8.6
```

(5)
```
    6 2
×   9.4
```

Review STEP 27

Calculate.

(1)
```
    2 6
×   1.7
```

(2)
```
    2 8
×   1.6
```

(3)
```
    6 2
×   1.8
```

1 Calculate.

6 points per question

Example

Count the digits from the right.

```
 1.④ …1
×0.③ …1
0.④② …2  (1+1=2)
```
↑
└── Add 0.

(1) First, calculate 14 × 3. ⟹
```
    1 4
×     3
    4 2
```

(2) Then, place the decimal point as many place values as the sum of the digits' decimals from the right.

(1)
```
    1.4
×   0.2
```
☐0.☐☐

(3)
```
    1.2
×   0.3
```

(5)
```
    1.8
×   0.4
```

(2)
```
    1.4
×   0.4
```

(4)
```
    1.6
×   0.2
```

STEP 1-5
Structure of
a Decimal

STEP 6-12
Addition of
Decimals

STEP 13-19
Subtraction of
Decimals

STEP 20-30
Multiplication
of Decimals

STEP 31-40
Decimal ÷
Whole Number

STEP 41-48
Decimal ÷
Decimal

2 Calculate.

5 points per question

(1)
```
    2 . 3
×   0 . 4
```

(3)
```
    2 . 6
×   0 . 3
```

(5)
```
    3 . 6
×   0 . 2
```

(2)
```
    2 . 4
×   0 . 4
```

(4)
```
    3 . 4
×   0 . 2
```

(6)
```
    4 . 2
×   0 . 2
```

3 Calculate.

5 points per question

(1)
```
    6 . 4
×   0 . 2
   ┌─┬─┬─┐
   │▯.▯│▯│
   └─┴─┴─┘
```

(4)
```
    7 . 4
×   0 . 4
```

(7)
```
    8 . 4
×   0 . 8
```

(2)
```
    6 . 3
×   0 . 3
```

(5)
```
    7 . 3
×   0 . 3
```

(8)
```
    8 . 6
×   0 . 9
```

(3)
```
    6 . 5
×   0 . 4
   ┌─┬─┬─┐
   │▯.▯│▯│
   └─┴─┴─┘
        ↑
   Remove 0 !
```

(6)
```
    6 . 3
×   0 . 7
```

Review STEP 28

Calculate.

(1)
```
   1.8
 × 0.3
```

(2)
```
   6.4
 × 0.7
```

(3)
```
   7.2
 × 0.6
```

1 Calculate.

4 points per question

Example

```
  3.4   ×10      34
× 4.2   ×10    × 4 2
  6 8            6 8
1 3 6          1 3 6
1 4.2 8  ÷100  1 4 2 8
```

Count the digits from the right
of the decimal points.

```
( 1 )
( 1 )
  ↓    1 + 1 = 2
( 2 )
```

(1)
```
   2.3
 × 2.6
```

(2)
```
   2.4
 × 3.8
```

(3)
```
   2.5
 × 4.5
```

2 Calculate.

8 points per question

(1)
$$\begin{array}{r} 3.6 \\ \times\ 5.8 \\ \hline \end{array}$$

(5)
$$\begin{array}{r} 4.2 \\ \times\ 6.7 \\ \hline \end{array}$$

(9)
$$\begin{array}{r} 6.4 \\ \times\ 7.3 \\ \hline \end{array}$$

(2)
$$\begin{array}{r} 4.7 \\ \times\ 5.3 \\ \hline \end{array}$$

(6)
$$\begin{array}{r} 4.6 \\ \times\ 6.4 \\ \hline \end{array}$$

(10)
$$\begin{array}{r} 6.8 \\ \times\ 8.3 \\ \hline \end{array}$$

(3)
$$\begin{array}{r} 4.6 \\ \times\ 7.1 \\ \hline \end{array}$$

(7)
$$\begin{array}{r} 3.4 \\ \times\ 7.9 \\ \hline \end{array}$$

(11)
$$\begin{array}{r} 6.7 \\ \times\ 8.4 \\ \hline \end{array}$$

(4)
$$\begin{array}{r} 3.4 \\ \times\ 8.3 \\ \hline \end{array}$$

(8)
$$\begin{array}{r} 4.8 \\ \times\ 6.7 \\ \hline \end{array}$$

Review STEP 29

Calculate.

(1)
```
    5.6
×   3.2
```

(2)
```
    4.8
×   6.4
```

(3)
```
    6.4
×   7.6
```

1 Calculate.

10 points per question

Example

The digits from the right
of the decimal points.

```
  1.3④ ⋯ 2
× 2.⑥ ⋯ 1
─────────
  8 0 4
2 6 8
─────────
3.④⑧④ ⋯ 3 (2 + 1 = 3)
```

First, calculate 134 × 26.
Then, carefully place the decimal point.

(1)
```
    1.3 4
×     2.8
┌─┬─┬─┬─┐
└─┴─┴─┴─┘
┌─┬─┬─┐
└─┴─┴─┘
┌─┬─┬─┬─┐
└─┴─┴─┴─┘
```

(2)
```
    1.6 8
×     2.6
```

2 **Calculate.**

8 points per question

(1)
```
   3.4 6
 ×   4.8
```

(2)
```
   3.4 6
 ×   5.2
```

(3)
```
   4.7 8
 ×   5.6
```

(4)
```
   5.6 4
 ×   5.4
```

(5)
```
   5.8 2
 ×   6.4
```

(6)
```
   6.7 2
 ×   6.2
```

(7)
```
   6.8 4
 ×   7.2
```

(8)
```
   7.6 8
 ×   6.8
```

(9)
```
   8.4 6
 ×   6.8
```

(10)
```
   9.5 8
 ×   6.7
```

Multiplication of Decimals

Date   /   /

Score   /100

Review STEP 20 STEP 21 Calculate.

4 points per question

(1) $0.2 \times 4 =$ ☐

(3) $0.4 \times 30 =$ ☐

(2) $3.4 \times 6 =$ ☐

(4) $3.4 \times 500 =$ ☐

Review STEP 22 – STEP 25 Calculate.

4 points per question

(1)
$$\begin{array}{r} 5.7 \\ \times\ \ 6 \\ \hline \end{array}$$

(3)
$$\begin{array}{r} 3.82 \\ \times\ \ \ \ 5 \\ \hline \end{array}$$

(5)
$$\begin{array}{r} 4.23 \\ \times\ \ 26 \\ \hline \end{array}$$

(2)
$$\begin{array}{r} 26.4 \\ \times\ \ \ \ 4 \\ \hline \end{array}$$

(4)
$$\begin{array}{r} 3.15 \\ \times\ \ 28 \\ \hline \end{array}$$

(6)
$$\begin{array}{r} 0.24 \\ \times\ \ 48 \\ \hline \end{array}$$

Review STEP 26 Calculate.

5 points per question

(1)
$$\begin{array}{r} 25 \\ \times\ 0.7 \\ \hline \end{array}$$

(3)
$$\begin{array}{r} 42 \\ \times\ 0.8 \\ \hline \end{array}$$

Review STEP 27 — Calculate.

5 points per question

(1)
```
    2 3
×   1.8
```

(4)
```
    5 2
×   2.8
```

Review STEP 28 STEP 29 — Calculate.

5 points per question

(1)
```
    1.4
×   0.3
```

(3)
```
    8.4
×   0.6
```

(5)
```
    4.2
×   5.6
```

(2)
```
    3.6
×   0.4
```

(4)
```
    2.4
×   4.8
```

(6)
```
    7.8
×   8.4
```

Review STEP 30 — Calculate.

5 points per question

(1)
```
    2.8 6
×     2.8
```

(2)
```
    5.8 2
×     6.2
```

71

Decimal ÷ Whole Number
Division of Decimals 1

Date / /

Score /100

Review STEP 30

Calculate.

(1)
```
   3.4 6
×    4.2
```

(2)
```
   6.7 2
×    6.4
```

1 Divide.

10 points per question

Example

$3.6 \div 3 = 1.2$

3.6 consists of thirty-six (36) 0.1s.

$36 \div 3 = 12$ so,

twelve (12) 0.1s.

$3.6 \div 3 = 1.2$

$\updownarrow$

$36 \div 3 = 12$

(1) $4.8 \div 4 = $ ☐

(3) $8.4 \div 2 = $ ☐

(2) $6.3 \div 3 = $ ☐

(4) $9.6 \div 3 = $ ☐

2 Divide.

6 points per question

(1) $10.4 \div 2 =$

(6) $16.8 \div 4 =$

(2) $12.8 \div 4 =$

(7) $12.6 \div 2 =$

(3) $12.6 \div 6 =$

(8) $15.9 \div 3 =$

(4) $12.6 \div 3 =$

(9) $20.8 \div 4 =$

(5) $15.5 \div 5 =$

(10) $24.8 \div 4 =$

Review STEP 31

Divide.

(1) $9.3 \div 3 = \boxed{}$

(2) $12.4 \div 2 = \boxed{}$

1 Calculate.

8 points per question

Example

$$
\begin{array}{r}
2 \\
3\overline{)7.2} \\
6 \\
\hline
1
\end{array}
$$

⇨

Place the decimal point.

$$
\begin{array}{r}
2. \\
3\overline{)7.2} \\
6 \\
\hline
1\;2
\end{array}
$$

⇨

$$
\begin{array}{r}
2.4 \\
3\overline{)7.2} \\
6 \\
\hline
1\;2 \\
1\;2 \\
\hline
0
\end{array}
$$

(1)

$$4\overline{)9.6}$$

(3)

$$3\overline{)5.1}$$

(5)

$$8\overline{)9.6}$$

(2)

$$7\overline{)8.4}$$

(4)

$$6\overline{)7.8}$$

2 Calculate.

5 points per question

(1)

$4\overline{)14.8}$

(2)

$3\overline{)16.5}$

(3)

$6\overline{)19.2}$

(4)

$7\overline{)22.4}$

(5)

$8\overline{)25.6}$

(6)

$9\overline{)37.8}$

(7)

$2\overline{)35.6}$

(8)

$5\overline{)74.5}$

(9)

$6\overline{)85.8}$

(10)

$7\overline{)85.4}$

(11)

$8\overline{)98.4}$

(12)

$4\overline{)78.8}$

Review STEP 32

Calculate.

(1)

$$5\overline{)9.5}$$

(2)

$$9\overline{)84.6}$$

---

1 **Divide.**

8 points per question

Example $0.6 \div 3 = 0.2$

0.6 consists of six (6) 0.1s.

When six (6) is divided by 3, $6 \div 3 = 2$

Then, two (2) 0.1s make 0.2

(1) $0.4 \div 2 =$ ☐

(4) $0.8 \div 4 =$ ☐

(2) $0.5 \div 5 =$ ☐

(5) $0.9 \div 3 =$ ☐

(3) $0.6 \div 2 =$ ☐

2 Calculate.

Example

$$9\overline{)4.5} \quad \Rightarrow \quad 9\overline{)4.5}^{\,0.} \quad \Rightarrow \quad 9\overline{)4.5}^{\,0.5}$$

Write 0 in the ones place of the quotient.

For the last example:
$$\begin{array}{r} 0.5 \\ 9\overline{)4.5} \\ \underline{4\ 5} \\ 0 \end{array}$$

(1) $4\overline{)1.2}$

(2) $6\overline{)1.2}$

(3) $7\overline{)1.4}$

(4) $8\overline{)1.6}$

(5) $3\overline{)1.8}$

(6) $9\overline{)1.8}$

(7) $6\overline{)2.4}$

(8) $7\overline{)3.5}$

(9) $5\overline{)4.5}$

(10) $8\overline{)7.2}$

(11) $7\overline{)6.3}$

(12) $9\overline{)8.1}$

Review STEP 33

Calculate.

(1)

7)6.3

(2)

5)3.5

---

1 Calculate.

10 points per question

Example 9.48 ÷ 4

```
       2.
   4)9.4 8
     8
     1 4
```
⇨
```
       2.3
   4)9.4 8
     8
     1 4
     1 2
     2 8
```
⇨
```
       2.3 7
   4)9.4 8
     8
     1 4
     1 2
     2 8
     2 8
       0
```

(1)

4)9.4 4

(2)

3)7.4 7

STEP 1-5
Structure of
a Decimal

STEP 6-12
Addition of
Decimals

STEP 13-19
Subtraction of
Decimals

STEP 20-30
Multiplication
of Decimals

STEP 31-40
Decimal ÷
Whole Number

STEP 41-48
Decimal ÷
Decimal

2 Calculate.

10 points per question

(1)

$$0.\boxed{0}\boxed{}$$
$$2\,)\overline{0.1\ 8}$$

(3)

$$4\,)\overline{0.3\ 2}$$

(2)

$$3\,)\overline{0.2\ 7}$$

(4)

$$6\,)\overline{0.4\ 2}$$

3 Calculate until there is no remainder.

10 points per question

(1)

$$\boxed{}.\boxed{}\boxed{}$$
$$4\,)\overline{9.4}$$

(2)

$$\boxed{}.\boxed{}\boxed{}$$
$$5\,)\overline{4.6}$$

(3)

$$8\,)\overline{10.8}$$

(4)

$$5\,)\overline{0.63}$$

Review STEP 34

Calculate.

(1)

$$8\overline{)0.48}$$

(2)

$$6\overline{)0.54}$$

1 Calculate until there is no remainder.

9 points per question

Example $16.5 \div 22$

$$
\begin{array}{r}
0.75 \\
22\overline{)16.5} \\
154 \\
\hline
110 \\
110 \\
\hline
0
\end{array}
$$

(1)

$$18\overline{)61.2}$$ $\boxed{}.\boxed{}$

(3)

$$15\overline{)12.9}$$ $0.$

(2)

$$23\overline{)52.9}$$

(4)

$$18\overline{)13.5}$$

2 Calculate until there is no remainder.

8 points per question

(1)

$15 \overline{)21.3}$

(2)

$15 \overline{)16.2}$

(3)

$18 \overline{)44.1}$

(4)

$15 \overline{)33.6}$

(5)

$32 \overline{)75.2}$

(6)

$26 \overline{)81.9}$

(7)

$25 \overline{)32.1}$ ☐.☐☐☐

(8)

$16 \overline{)10.8}$

81

Decimal ÷ Whole Number
Find Quotient as Round Number 1

Date / /

Score /100

Review STEP 35

Calculate until there is no remainder.

(1)

$$18\overline{)18.9}$$

(2)

$$16\overline{)13.6}$$

---

1 **Find the answer by rounding the quotient, so it is a single digit.**

10 points per question

Example Find the first digit.

$$
\begin{array}{r}
\overset{2}{1.5} \\
6\overline{)9.2} \\
6 \\
\hline
3\,2 \\
3\,0 \\
\hline
2
\end{array}
$$

Round the first two digits of the quotient.
Then, find the answer "2" as a single digit.

Round off.

$$
\begin{array}{r}
\overset{5}{0.4\,6} \\
6\overline{)2.8} \\
2\,4 \\
\hline
4\,0 \\
3\,6 \\
\hline
4
\end{array}
$$

Round the first two digits of the quotient.
Then, find the answer "0.5" as a single digit.

(1)

$$7\overline{)1.2}$$

(2)

$$8\overline{)35.4}$$

2 Find the answer by rounding the quotient, so it is a single digit.

10 points per question

(1)

$21\overline{)13.3}$

(5)

$29\overline{)3.14}$

(2)

$31\overline{)17.5}$

(6)

$44\overline{)7.73}$

(3)

$13\overline{)29.8}$

(7)

$60\overline{)12.5}$

(4)

$53\overline{)88.2}$

(8)

$57\overline{)2.746}$

Review STEP 36

Find the answer by rounding the quotient, so it is a single digit.

(1)

$$7 \overline{)2.3}$$

(2)

$$54 \overline{)16.8}$$

1   **Find the answer by rounding the quotient to the first two digits.**

10 points per question

Example   Find the first two digits.

$$
\begin{array}{r}
0.6\overset{8}{\cancel{7}6} \\
21 \overline{)14.2} \\
126 \\
\hline
160 \\
147 \\
\hline
130 \\
126 \\
\hline
4
\end{array}
$$

← Use the first three digit to round the quotient to the first two digits.
Then, find the answer "0.68".

The third digit of the quotient is "6". It is in the thousandths place.

(1)

$$9 \overline{)1.2}$$

(2)

$$7 \overline{)1.6}$$

2 Find the answer by rounding the quotient to the first two digits.

10 points per question

(1)

$19\overline{)13.5}$

(2)

$34\overline{)23.3}$

(3)

$13\overline{)27.2}$

(4)

$53\overline{)65.1}$

(5)

$28\overline{)31.3}$

(6)

$53\overline{)57.6}$

(7)

$61\overline{)12.58}$

(8)

$53\overline{)23.456}$

STEP **38**

Decimal ÷ Whole Number

Division of Decimals with Remainder 1

Date / /

Score

/100

Review STEP 37

Find the answer by rounding the quotient to the first two digits.

(1)

$$28\overline{)32.3}$$

(2)

$$61\overline{)12.68}$$

---

1 **Find the quotient in the ones place, and write the remainder.**

5 points per question

Example $16.7 \div 3$

$$16.7 \div 3 = 5 \text{ R } 1.7$$

"R" means the "remainder."

$$
\begin{array}{r}
5 \\
3\overline{)16.7} \\
15 \\
\hline
1.7
\end{array}
$$

remainder → 1.7

Place the decimal point of the remainder to align with the decimal point of the dividend.

(1)

$$3\overline{)7.2}$$

(3)

$$4\overline{)13.3}$$

(2)

$$5\overline{)14.7}$$

(4)

$$6\overline{)38.5}$$

2 Find the quotient in the ones place, and write the remainder.

8 points per question

(1)

$3 \overline{)12.4}$

(2)

$4 \overline{)32.3}$

(3)

$7 \overline{)97.4}$

(4)

$8 \overline{)70.5}$

(5)

$6 \overline{)132.8}$

(6)

$9 \overline{)108.1}$

(7)

$12 \overline{)45.2}$

(8)

$26 \overline{)74.8}$

(9)

$23 \overline{)50.3}$

(10)

$16 \overline{)35.8}$

Decimal ÷ Whole Number

Whole Number ÷ Whole Number 1

Date / /

Score /100

Review STEP 38

Find the quotient in the ones place, and write the remainder.

(1)

$6\overline{)26.1}$

(2)

$8\overline{)35.6}$

1 **Calculate until there is no remainder.**

8 points per question

Example

$12 \div 5$

$$5\overline{)12} \begin{array}{r} 2.4 \\ \underline{10} \\ 2\,0 \\ \underline{2\,0} \\ 0 \end{array}$$

(2)

$5\overline{)14}$

(4)

$5\overline{)17}$

(1)

$5\overline{)13}$

(3)

$5\overline{)16}$

(5)

$5\overline{)18}$

2 Calculate until there is no remainder.

5 points per question

(1)

$4 \overline{) 6}$

(5)

$4 \overline{) 10}$

(9)

$4 \overline{) 38}$

(2)

$4 \overline{) 14}$

(6)

$4 \overline{) 22}$

(10)

$8 \overline{) 52}$

(3)

$6 \overline{) 15}$

(7)

$6 \overline{) 27}$

(11)

$5 \overline{) 48}$

(4)

$8 \overline{) 20}$

(8)

$5 \overline{) 24}$

(12)

$8 \overline{) 60}$

Review STEP 39

Calculate until there is no remainder.

(1)

$$5\overline{)23}$$

(2)

$$4\overline{)38}$$

1 Calculate until there is no remainder.

4 points per question

Example 6 ÷ 15

$$
\begin{array}{r}
0.4 \\
15\overline{)6.0} \\
60 \\
\hline
0
\end{array}
$$

— Write 0 because it cannot be divided, and place the decimal point here.

31 ÷ 25

$$
\begin{array}{r}
1.24 \\
25\overline{)31.00} \\
25 \\
\hline
60 \\
50 \\
\hline
100 \\
100 \\
\hline
0
\end{array}
$$

(1)

$$
\begin{array}{r}
0. \\
12\overline{)6.0}
\end{array}
$$

(2)

$$15\overline{)9}$$

(3)

$$14\overline{)7}$$

2 Calculate until there is no remainder.

11 points per question

(1)

$15\overline{)12}$

(2)

$26\overline{)13}$

(3)

$38\overline{)19}$

(4)

$35\overline{)21}$

(5)

$16\overline{)40}$

(6)

$25\overline{)12}$

(7)

$25\overline{)52}$

(8)

$48\overline{)60}$

Decimal ÷ Whole Number

Date / /

Score /100

Review STEP 31 — Calculate.

5 points per question

(1) $15.3 \div 3 =$ ☐

(2) $12.8 \div 2 =$ ☐

Review STEP 32 STEP 33 — Calculate.

5 points per question

(1)

$4 \overline{)9.2}$

(2)

$9 \overline{)7.2}$

Review STEP 34 — Calculate until there is no remainder.

6 points per question

(1)

$3 \overline{)6.42}$

(2)

$5 \overline{)0.62}$

Review STEP 35 — Calculate until there is no remainder.

6 points per question

(1)

$22 \overline{)16.5}$

(2)

$38 \overline{)89.3}$

Review STEP 36

Find the answer by rounding the quotient, so it is a single digit.

7 points per question

(1)

$21 \overline{)14.5}$

(2)

$54 \overline{)5.73}$

Review STEP 37

Find the answer by rounding the quotient to the first two digits.

7 points per question

(1)

$24 \overline{)50.3}$

(2)

$53 \overline{)57.7}$

Review STEP 38

Find the quotient to the ones place, and write the remainder.

7 points per question

(1)

$3 \overline{)46.7}$

(2)

$12 \overline{)45.1}$

Review STEP 39 40

Calculate until there is no remainder. *7 points per question*

(1)

$4 \overline{)5}$

(2)

$12 \overline{)42}$

Review STEP 40

Calculate until there is no remainder.

(1)

$15\overline{)18}$

(2)

$48\overline{)36}$

1 Calculate.

5 points per question

Example $48 \div 2.4$

$2.4\overline{)48}$

×10 ↓   ↓ ×10

$24\overline{)480}$

$$\begin{array}{r} 2\ 0 \\ 2.4\overline{)4\ 8\ 0} \\ 4\ 8 \\ \hline 0 \end{array}$$

Remove 0, and add 0 to the dividend (48). Then, calculate 480 ÷ 24.

In division, the quotient is not changed if both the divisor and the dividend are multiplied by 10.

$10 \div 2 = 5$
↓×10 ↓ ×10
$100 \div 20 = 5$
↓×10 ↓ ×10
$1000 \div 200 = 5$

(1)

$1.4\overline{)28}$

(2)

$2.4\overline{)72}$

(3)

$3.2\overline{)64}$

(4)

$4.3\overline{)86}$

2 Calculate.

8 points per question

(1)

$$3.5 \overline{)42}$$

(2)

$$2.6 \overline{)91}$$

(3)

$$3.5 \overline{)98}$$

(4)

$$1.5 \overline{)12}$$

(5)

$$1.6 \overline{)40}$$

(6)

$$2.5 \overline{)200}$$

(7)

$$2.5 \overline{)375}$$

(8)

$$2.4 \overline{)120}$$

(9)

$$6.5 \overline{)325}$$

(10)

$$8.5 \overline{)425}$$

Review STEP 41

Calculate.

(1)

$4.2 \overline{)84}$

(2)

$2.5 \overline{)150}$

1 Calculate.

10 points per question

Example $7.8 \div 6.5$

Place the decimal point of the quotient to align with the moved one.

 $\Rightarrow$ $\Rightarrow$

$$
\begin{array}{r}
1.2 \\
6.5 \overline{)7.8.0} \\
6\ 5 \\
\hline
1\ 3\ 0 \\
1\ 3\ 0 \\
\hline
0
\end{array}
$$

Multiply the divisor by 10 so that it can be a whole number. Then, move the decimal point to right.

Multiply the dividend by 10, and move the decimal point to right as well.

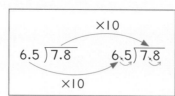

(1)

$1.4 \overline{)2.8}$

(2)

$3.2 \overline{)9.6}$

2 Calculate until there is no remainder.

8 points per question

(1)

$$3.5 \overline{)4.2}$$

(2)

$$6.2 \overline{)9.3}$$

(3)

$$1.8 \overline{)4.5}$$

(4)

$$2.5 \overline{)8.5}$$

(5)

$$1.5 \overline{)4.8}$$

(6)

$$7.3 \overline{)58.4}$$

(7)

$$4.2 \overline{)10.5}$$

(8)

$$2.8 \overline{)23.8}$$

(9)

$$3.5 \overline{)2.1}$$

(10)

$$4.8 \overline{)2.4}$$

Review STEP 42

Calculate until there is no remainder.

(1)

$1.4\overline{)4.2}$

(2)

$2.4\overline{)8.4}$

1 Calculate.

10 points per question

Example $1.56 \div 1.2$

$1.2\overline{)1.56}$ ×10 ×10 $\Rightarrow$

$$
\begin{array}{r}
1.3 \\
1.2\overline{)1.5.6} \\
\underline{1\ 2} \\
3\ 6 \\
\underline{3\ 6} \\
0
\end{array}
$$

(1)

$1.6\overline{)2.72}$

(2)

$2.1\overline{)7.14}$

2 Calculate until there is no remainder.

10 points per question

(1)

$$2.5 \overline{)5.75}$$

(2)

$$2.7 \overline{)2.16}$$

(3)

$$3.2 \overline{)13.76}$$

(4)

$$5.6 \overline{)3.64}$$

(5)

$$1.8 \overline{)2.25}$$

(6)

$$3.5 \overline{)7.98}$$

(7)

$$2.5 \overline{)1.35}$$

(8)

$$3.2 \overline{)4.64}$$
(0.)

99

Decimal ÷ Decimal

Division: ÷ Decimal 3

Date / /

Score /100

Review STEP **42**

Calculate until there is no remainder.

(1)

$1.2 \overline{)2.76}$

(2)

$3.5 \overline{)1.82}$

1 Calculate until there is no remainder.

9 points per question

Example $4.5 \div 0.75$

Add 0.

$0.75 \overline{)4.50} \leftarrow$
×100   ×100

Multiply both the divisor and
the dividend by 100.

$$0.75 \overline{)4.50} \begin{array}{r} 6 \\ \hline \end{array}$$
$$\underline{450}$$
$$0$$

The quotient becomes larger than the dividend (4.5)
when it is divided by a number smaller than 1.

(1)

$0.7 \overline{)2.1}$

(3)

$0.24 \overline{)1.2}$

(2)

$0.8 \overline{)1.2}$

(4)

$0.36 \overline{)1.8}$

2 Calculate until there is no remainder.

8 points per question

(1)

$0.03 \overline{)0.09}$

(2)

$0.05 \overline{)0.4}$

(3)

$0.75 \overline{)0.9}$

(4)

$0.31 \overline{)12.4}$

(5)

$0.8 \overline{)0.76}$

(6)

$0.5 \overline{)1.37}$

(7)

$0.65 \overline{)2.34}$

(8)

$0.86 \overline{)3.87}$

STEP 45

Decimal ÷ Decimal

Division: ÷ Decimal 4

Date / /

Score /100

Review STEP 44

Calculate until there is no remainder.

(1) $0.62\overline{)3.1}$

(2) $0.24\overline{)0.84}$

1 Calculate.

10 points per question

Example 0.598 ÷ 0.26

$0.26\overline{)0.598}$ ⇨

$$\begin{array}{r} 2.3 \\ 0.26\overline{)0.59.8} \\ 52 \\ \hline 7\,8 \\ 7\,8 \\ \hline 0 \end{array}$$

(1) $0.74\overline{)0.888}$

(2) $0.35\overline{)0.875}$

2 Calculate until there is no remainder.

10 points per question

(1)

$1.24 \overline{)2.604}$

(2)

$1.45 \overline{)1.885}$

(3)

$2.24 \overline{)3.136}$

(4)

$3.14 \overline{)7.222}$

(5)

$3.24 \overline{)7.938}$

(6)

$4.38 \overline{)5.475}$

(7)

$4.25 \overline{)5.27}$

(8)

$5.65 \overline{)11.978}$

Review STEP **45**

Calculate.

(1)

$$0.24 \overline{)0.744}$$

(2)

$$1.34 \overline{)5.628}$$

1 Find the answer by rounding the quotient to the nearest tenths place.

10 points per question

Example   Find the quotient of $5.2 \div 0.6$ by rounding to the nearest tenths place.

$$0.6 \overline{)5.2}$$ ⟹

$$
\begin{array}{r}
8.6\,6 \\
0.6 \overline{)5.2} \\
4\,8 \\
\hline
4\,0 \\
3\,6 \\
\hline
4\,0 \\
3\,6 \\
\hline
4
\end{array}
$$

← Round to the nearest tenths place. Then, find the answer "8.7" here.

(1)

$$4.7 \overline{)7.5}$$

(2)

$$2.6 \overline{)8.7}$$

2 Find the answer by rounding the quotient to the nearest tenths place.

10 points per question

(1)

$$2.6 \overline{)9}$$

(2)

$$0.9 \overline{)75}$$

(3)

$$9.6 \overline{)25}$$

(4)

$$3.7 \overline{)33.6}$$

(5)

$$2.9 \overline{)8.3}$$

(6)

$$6.7 \overline{)21.3}$$

(7)

$$8.9 \overline{)24.2}$$

(8)

$$0.26 \overline{)0.48}$$

Review STEP **46**

Find the answer by rounding the quotient to the nearest tenths place.

(1)

$$2.6\overline{)8.6}$$

(2)

$$1.28\overline{)1.82}$$

---

1 **Find the answer by rounding the quotient to the nearest hundredths place.**

20 points

Example

Find the quotient of 0.673 ÷ 0.26 by rounding to the nearest hundredths place.

$$0.36\overline{)0.875}$$

```
                 9
          2.5 8 8    ← Round to the nearest
  0.26 ) 0.6 7.3         hundredths place.
         5 2            Then, find the answer
         1 5 3          "2.59" here.
         1 3 0
           2 3 0
           2 0 8
             2 2 0
             2 0 8
               1 2
```

2 Find the answer by rounding the quotient to the nearest hundredths place.

20 points per question

(1)

$$1.34 \overline{)0.73}$$

(3)

$$6.57 \overline{)8.94}$$

(2)

$$2.56 \overline{)4.674}$$

(4)

$$0.09 \overline{)3.553}$$

Review STEP 47

Find the answer by rounding the quotient to the nearest hundredths place.

(1)

$0.24\overline{)0.76}$

(2)

$1.36\overline{)2.573}$

---

1 **Find the quotient to the first two digits, and write the remainder.**

10 points per question

Example $6.7 \div 1.4 = 4.7 \text{ R } 0.12$

```
        4.7
   1.4 )6.7
        5 6
        1 1 0
          9 8
        0.1 2
```

Place the decimal point of the remainder so it aligns with the original place of that of the dividend.

(1)

$1.3\overline{)9.6}$

(2)

$3.2\overline{)10.9}$

2 Find the quotient to the first two digits, and write the remainder.

10 points per question

(1)

$0.9 \overline{)0.89}$

(2)

$6.5 \overline{)0.89}$

(3)

$7.2 \overline{)22.74}$

(4)

$7.3 \overline{)30.78}$

(5)

$1.24 \overline{)2.616}$

(6)

$0.12 \overline{)0.287}$

(7)

$2.46 \overline{)5.925}$

(8)

$3.82 \overline{)11.854}$

Decimal ÷ Decimal

Date / /

Score /100

Review STEP **41**   **Calculate.**

7 points per question

(1)

$2.4\overline{)48}$

(2)

$3.5\overline{)98}$

Review STEP **42**   **Calculate until there is no remainder.**

7 points per question

(1)

$1.3\overline{)7.8}$

(2)

$3.5\overline{)9.8}$

Review STEP **43**   **Calculate until there is no remainder.**

7 points per question

(1)

$1.2\overline{)2.16}$

(2)

$2.5\overline{)4.05}$

Review STEP **44**   **Calculate until there is no remainder.**

7 points per question

(1)

$0.94\overline{)4.7}$

(2)

$0.42\overline{)1.47}$

Review STEP 45   Calculate until there is no remainder.

7 points per question

(1)

$0.24\overline{)0.516}$

(2)

$3.25\overline{)4.745}$

Review STEP 46 STEP 47   Find the answer by rounding the quotient, so it is a single digit.

7 points per question

(1)

$3.2\overline{)6.8}$

(2)

$4.08\overline{)5.76}$

Review STEP 48   Find the quotient to the first two digits, and write the remainder.

8 points per question

(1)

$1.4\overline{)9.5}$

(2)

$1.34\overline{)2.826}$

Math Boosters

Grades 3-5 Decimals

Answer Key

STEP 1 (P.4 · 5)

❶
(1) 0.2   (4) 0.3
(2) 0.4   (5) 0.9
(3) 0.6   (6) 0.7

❷
(1) 1.1   (3) 2.6
(2) 1.3   (4) 2.9

❸
(1) 0.5   (4) 7.2
(2) 0.8   (5) 6.8
(3) 4.5   (6) 10.7

STEP 2 (P.6 · 7)

■ Review of Step 1
(1) 0.6   (2) 1.4

❶
(1) 6(cm) 4(mm)   (3) 0.4
(2) 0.1   (4) 6.4

❷
A 0.8   C 8.2
B 4   D 11.6

❸
A 0.6   B 0.8   C 1.1

❹
(1) 0.6   (4) 5.4   (7) 0.1
(2) 1   (5) 0.1   (8) 1.7
(3) 1.8   (6) 1.4

STEP 3 (P.8 · 9)

■ Review of Step 2
9.5

❶
(1) A 0.7   (2) A 1.91
     B 2.1   B 1.98
     C 3.9   C 2.05

❷

0   1   2   3   4
↑0.4   ↑1.3   ↑2.8   ↑3.3

❸
(1) 1.7   (2) 1.1   (3) 3.1

❹ 0 < 0.01 < 0.08 < 0.12

❺
(1) <   (2) >   (3) >

STEP 4 (P.10 · 11)

■ Review of Step 3
A 0.2   B 1.6   C 2.8

❶
(1) 1.1   (4) 1.4   (7) 2
(2) 3.2   (5) 1.7   (8) 8
(3) 0.9   (6) 3.3

❷
(1) 6   (4) 15   (7) 33
(2) 9   (5) 20   (8) 112
(3) 10   (6) 21

❸
(1) 4   (4) 23
(2) 9   (5) 30
(3) 12   (6) 125

❹
(1) 256.7   (4) 0.2567   (7) 0.06
(2) 2567   (5) 6
(3) 2.567   (6) 0.4

STEP 5 (P.12 · 13)

■ Review of Step 4
(1) 0.7   (2) 0.3   (3) 0.17

❶ 1.325L

❷
(1) 4.375   (3) 0.037
(2) 25.748

③ A 3.408 C 3.477
B 3.43

④ (1) 15.2 (3) 0.0538
(2) 152 (4) 0.00538

TEST
(P.14 · 15)

Review of Step 1
(1) 2.26 (2) 3.04

Review of Step 2
7.6

Review of Step 3
A 3.93 B 3.99 C 4.05

Review of Step 4,5
(1) 6.45 (4) 360.4 (7) 0.063
(2) 17.342 (5) 16.3 (8) 0.005
(3) 0.15 (6) 0.542

STEP 6
(P.16 · 17)

■ Review of Step 5
(1) 8.5 (2) 0.032

① (1) 0.6 (3) 0.4 (5) 0.9
(2) 0.8 (4) 0.8 (6) 0.5

② (1) 1.6 (6) 1.8 (11) 2.8
(2) 1.8 (7) 2.7 (12) 5.7
(3) 2.6 (8) 3.9 (13) 3.9
(4) 3.8 (9) 2.7 (14) 2.8
(5) 1.8 (10) 4.7

STEP 7
(P.18 · 19)

■ Review of Step 6
(1) 0.7 (3) 0.9
(2) 2.9 (4) 4.6

① (1) 1 (4) 2 (7) 3
(2) 1 (5) 3
(3) 2 (6) 2

② (1) 1.3 (4) 1.2 (7) 4.3
(2) 1.4 (5) 3.4 (8) 4.6
(3) 1.6 (6) 4.6

③ (1) 1.9 (3) 5.7 (5) 5.6
(2) 2.8 (4) 7.5

STEP 8
(P.20 · 21)

■ Review of Step 7
(1) 1 (3) 4.4
(2) 3 (4) 5.8

① (1) 2.6 (4) 2.9 (7) 2.4
(2) 2.7 (5) 2.8
(3) 2.8 (6) 2.7

② (1) 3.8 (4) 3.8 (7) 4.9
(2) 3.9 (5) 4.9 (8) 4.9
(3) 3.7 (6) 4.7

③ (1) 5.5 (3) 5.7 (5) 8.9
(2) 4.7 (4) 7.4

STEP 9
(P.22 · 23)

■ Review of Step 8
(1) 2.7 (3) 4.9
(2) 3.9 (4) 7.8

① (1) 1.1 (4) 1.5 (7) 1.2
(2) 1.3 (5) 1.5
(3) 1.2 (6) 1.2

② (1) 2.1 (4) 2.2 (7) 2.2
(2) 2.2 (5) 2.3 (8) 2.1
(3) 2.2 (6) 2.3

❸ (1) 3.4   (3) 3.3   (5) 3.1
     (2) 3.2   (4) 3.1

STEP 10
(P.24 · 25)

■ Review of Step 9
(1) 1.5   (3) 2.3
(2) 2.4   (4) 1.4

❶ (1) 6.1   (3) 6.1   (5) 6.2
     (2) 6.1   (4) 6.1   (6) 6.2

❷ (1) 7.3   (4) 7.5   (7) 7.1
     (2) 7.3   (5) 7.2   (8) 7.2
     (3) 7.2   (6) 7.2

❸ (1) 8.2   (3) 8.1   (5) 9.1
     (2) 9.5   (4) 9.4   (6) 8.1

STEP 11
(P.26 · 27)

■ Review of Step 10
(1) 6.1   (3) 7.2
(2) 7.2   (4) 8.1

❶

(1)
$$\begin{array}{r} 0.4 \\ +0.35 \\ \hline 0.75 \end{array}$$

(3)
$$\begin{array}{r} 15.3 \\ + 1.24 \\ \hline 16.54 \end{array}$$

(2)
$$\begin{array}{r} 1.68 \\ +1.3 \\ \hline 2.98 \end{array}$$

(4)
$$\begin{array}{r} 0.73 \\ +21.2 \\ \hline 21.93 \end{array}$$

❷ (1) 10.64   (4) 26.15
     (2) 3.17   (5) 8.34
     (3) 18.23   (6) 19.18

❸ (1) 4.98   (3) 9.23
     (2) 4.99   (4) 6

STEP 12
(P.28 · 29)

■ Review of Step 11
(1) 0.95   (2) 16.25

❶

(1)
$$\begin{array}{r} 0.2 \\ +0.765 \\ \hline 0.965 \end{array}$$

(3)
$$\begin{array}{r} 0.04 \\ +5.832 \\ \hline 5.872 \end{array}$$

(2)
$$\begin{array}{r} 1.634 \\ +0.24 \\ \hline 1.874 \end{array}$$

(4)
$$\begin{array}{r} 4.764 \\ +2.23 \\ \hline 6.994 \end{array}$$

❷ (1) 1.118   (4) 18.014
     (2) 9.082   (5) 8.017
     (3) 6.499   (6) 10.203

❸ (1) 3.679   (3) 7.197
     (2) 46.699   (4) 20.201

TEST
(P.30 · 31)

Review of Step 6
(1) 0.8   (3) 4.9
(2) 1.6   (4) 0.8

Review of Step 7
(1) 2   (3) 4.4   (5) 4.8
(2) 4.7   (4) 3   (6) 2

Review of Step 8
(1) 2.8   (3) 5.9
(2) 3.9   (4) 5.8

Review of Step 9
(1) 1.1   (3) 2.2
(2) 2.1   (4) 2.2

Review of Step 10
(1) 6.1   (3) 7.1
(2) 6.1   (4) 8.2

Review of Step 11,12

(1)
```
  0.4
+0.35
─────
 0.75
```

(3)
```
  3.847
+1.754
──────
 5.601
```

(2)
```
   3.75
+12.6
──────
 16.35
```

(4)
```
  15.632
+  8.376
───────
 24.008
```

STEP 13

■ Review of Step 6

(1) 1.5		(3) 1.8	
(2) 3.7		(4) 1.9	

❶
(1) 0.2	(3) 0.6	(5) 0.2
(2) 0.3	(4) 0.3	(6) 0.1

❷
(1) 1.1	(6) 2.1	(11) 4.5
(2) 1.3	(7) 2.3	(12) 4.1
(3) 1.2	(8) 2.1	(13) 5.1
(4) 2.1	(9) 3.3	(14) 5.3
(5) 2.2	(10) 3.3	

STEP 14

■ Review of Step 13

(1) 0.2		(3) 1.2	
(2) 0.5		(4) 3.1	

❶
(1) 1	(3) 2	(5) 2
(2) 1	(4) 3	(6) 2

❷
(1) 0.3	(6) 1.2	(11) 3.8
(2) 0.6	(7) 2.7	(12) 4.6
(3) 0.5	(8) 2.7	(13) 5.7
(4) 0.4	(9) 2.6	(14) 6.9
(5) 1.8	(10) 2.8	

STEP 15

■ Review of Step 14

(1) 2		(3) 4.4	
(2) 2		(4) 0.5	

❶
(1) 0.5	(3) 0.8	(5) 0.8
(2) 0.8	(4) 0.8	(6) 0.5

❷
(1) 1.8	(6) 2.6	(11) 3.9
(2) 1.6	(7) 2.9	(12) 3.5
(3) 1.7	(8) 2.3	(13) 4.3
(4) 1.1	(9) 3.5	(14) 4.8
(5) 2.7	(10) 3.9	

STEP 16

■ Review of Step 15

(1) 0.8		(3) 2.9	
(2) 1.7		(4) 4.8	

❶
(1) 0.2	(3) 0.6	(5) 0.6
(2) 0.1	(4) 0.1	(6) 0.6

❷
(1) 1.4	(5) 2.4	(9) 1.8
(2) 0.5	(6) 0.2	(10) 5.2
(3) 1.4	(7) 0.3	
(4) 0.1	(8) 2.2	

❸
(1) 12.2	(3) 4.2
(2) 12	(4) 6.6

STEP 17

■ Review of Step 16

(1) 0.4	(2) 1.3	(3) 1.5

❶
(1) 3.8	(3) 1.9	(5) 2.8
(2) 2.8	(4) 3.8	(6) 1.9

❷ (1) 1.3　　(5) 0.5　　(9) 0.7
(2) 2.3　　(6) 0.4　　(10) 0.2
(3) 1.4　　(7) 0.8
(4) 2.4　　(8) 0.3

❸ (1) 1.3　　(3) 0.5
(2) 2.8　　(4) 0.4

STEP 18　　　　　　　　　(P.42・43)

■ Review of Step 17
(1) 3.9　　　　(2) 2.9

❶
(1)
$$\begin{array}{r} 7.84 \\ -3.52 \\ \hline 4.32 \end{array}$$
(3)
$$\begin{array}{r} 4.56 \\ -2.31 \\ \hline 2.25 \end{array}$$

(2)
$$\begin{array}{r} 8.46 \\ -2.3 \\ \hline 6.16 \end{array}$$
(4)
$$\begin{array}{r} 0.17 \\ -0.04 \\ \hline 0.13 \end{array}$$

❷ (1) 3.24　　(5) 1.86
(2) 4.71　　(6) 6.89
(3) 5.96　　(7) 0.31
(4) 0.78　　(8) 2.95

❸ (1) 2.46　　(3) 1.94
(2) 2.92　　(4) 2.58

STEP 19　　　　　　　　　(P.44・45)

■ Review of Step 18
(1) 4.12　　　　(2) 6.06

❶
(1)
$$\begin{array}{r} 7.856 \\ -5.632 \\ \hline 2.224 \end{array}$$
(3)
$$\begin{array}{r} 7.865 \\ -5.75 \\ \hline 2.115 \end{array}$$

(2)
$$\begin{array}{r} 6.584 \\ -5.463 \\ \hline 1.121 \end{array}$$
(4)
$$\begin{array}{r} 8.437 \\ -8.4 \\ \hline 0.037 \end{array}$$

❷ (1) 1.362　　(5) 0.982
(2) 1.891　　(6) 0.038
(3) 1.484　　(7) 0.891
(4) 0.872　　(8) 5.679

❸ (1) 1.124　　(3) 0.881
(2) 0.985　　(4) 0.236

TEST　　　　　　　　　(P.46・47)

Review of Step 13
(1) 0.2　　(3) 5.1
(2) 0.3　　(4) 4.2

Review of Step 14
(1) 1　　　(4) 3.7
(2) 2　　　(5) 0.8
(3) 0.3

Review of Step 15
(1) 0.4　　(3) 4.9
(2) 0.3　　(4) 6.3

Review of Step 16,17
(1) 0.4　　(3) 1.8
(2) 1.7　　(4) 0.7

Review of Step 18
(1)
$$\begin{array}{r} 7.65 \\ -3.42 \\ \hline 4.23 \end{array}$$
(2)
$$\begin{array}{r} 6.23 \\ -2.5 \\ \hline 3.73 \end{array}$$
(3)
$$\begin{array}{r} 3.4 \\ -1.26 \\ \hline 2.14 \end{array}$$

Review of Step 19
(1)
$$\begin{array}{r} 7.86 \\ -5.953 \\ \hline 1.907 \end{array}$$
(3)
$$\begin{array}{r} 4.3 \\ -3.547 \\ \hline 0.753 \end{array}$$

(2)
$$\begin{array}{r} 6.258 \\ -4.68 \\ \hline 1.578 \end{array}$$

STEP 20

(P.48 · 49)

■ Review of Step 6
(1) 3.9   (3) 4.7
(2) 2.7   (4) 2.8

❶
(1) 0.8   (4) 4.8
(2) 3.2   (5) 2
(3) 2.4

❷
(1) 2.4   (5) 2.8   (9) 6.4
(2) 3.6   (6) 4.2   (10) 9.6
(3) 4.8   (7) 4.6
(4) 6   (8) 6.9

❸
(1) 12.4   (3) 12.8   (5) 12.8
(2) 12.6   (4) 10.5

STEP 21

(P.50 · 51)

■ Review of Step 20
(1) 5.6   (3) 15.5
(2) 6.9   (4) 24.8

❶
(1) 9   (3) 8   (5) 10
(2) 60   (4) 60

❷
(1) 20   (5) 46   (9) 24
(2) 24   (6) 69   (10) 240
(3) 24   (7) 36
(4) 48   (8) 360

❸
(1) 240   (4) 690
(2) 480   (5) 3600
(3) 240

STEP 22

(P.52 · 53)

■ Review of Step 21
(1) 12   (3) 300
(2) 80   (4) 1400

❶
(1)
$$\begin{array}{r} 0.7 \\ \times \quad 4 \\ \hline 2.8 \end{array}$$
(3) 7.2
(4) 5.2
(5) 7.2

(2)
$$\begin{array}{r} 0.9 \\ \times \quad 7 \\ \hline 6.3 \end{array}$$

❷
(1) 10.4   (6) 34.8
(2) 17.4   (7) 24.8
(3) 29.4   (8) 64.2
(4) 16.2
(5)
$$\begin{array}{r} 3.5 \\ \times \quad 4 \\ \hline 14.0 \end{array}$$

❸
(1) 21.6   (5)
$$\begin{array}{r} 42.5 \\ \times \quad 6 \\ \hline 255.0 \end{array}$$
(2) 31.2
(3) 25.2
(4) 245.6

STEP 23

(P.54 · 55)

■ Review of Step 22
(1) 21.5   (2) 22.4   (3) 98.4

❶
(1)
$$\begin{array}{r} 1.23 \\ \times \quad 2 \\ \hline 2.46 \end{array}$$
(2) 3.72
(3) 8.56
(4) 9.78
(5) 4.32

❷
(1) 2.25   (5) 21.42
(2) 3.84   (6) 20.15
(3) 8.28   (7) 5.12
(4) 16.92   (8) 6.4

❸
(1) 16.3   (4) 23.04
(2) 19.1   (5) 31.92
(3) 24.66

STEP 24 (P.56 · 57)

■ Review of Step 23
(1) 3.84　　　　(3) 22.92
(2) 4.28

❶ (1) 17.68　　(4) 24.96
(2) 19.88　　(5) 36.21
(3) 19.24

❷ (1) 53.25　　(5) 81.9
(2) 37.44　　(6) 47.04
(3) 82.62　　(7) 110.16
(4) 62.44　　(8) 191.14

❸ (1) 92.88　　(4) 291.48
(2) 57.04　　(5) 329.94
(3) 148.07

STEP 25 (P.58 · 59)

■ Review of Step 24
(1) 16.32　　　　(3) 101.52
(2) 80.75

❶ (1) 5.04　　(3) 5.12　　(5) 9.24
(2) 3.64　　(4) 4.08

❷ (1) 10.08　　(11) 19.95
(2) 12.48　　(12) 22.26
(3) 12.88　　(13) 13.64
(4) 11.28　　(14) 28.8
(5) 22.08　　(15) 60.48
(6) 13.52
(7) 14.72
(8) 18.25
(9) 20.16
(10) 11.55

STEP 26 (P.60 · 61)

■ Review of Step 25
(1) 4.32　　　　(3) 29.14
(2) 16.32

❶ (1) 7.2　　(3) 9.6　　(5) 5.4
(2) 8.4　　(4) 6.4

❷ (1) 17.5　　(4) 14.8　　(7) 16.2
(2) 13.6　　(5) 29.4　　(8) 16.8
(3) 14.4　　(6) 21.5

❸ (1) 12.6　　(3) 22.5　　(5) 20.8
(2) 10　　(4) 11.6

STEP 27 (P.62 · 63)

■ Review of Step 26
(1) 7.2　　　(2) 12.6　　　(3) 17.2

❶
(1)　16
　×1.2
　　32
　16
　19.2

(3)　26
　×1.3
　　78
　26
　33.8

(5)　34
　×1.6
　204
　34
　54.4

(2)　17
　×1.3
　　51
　17
　22.1

(4)　23
　×1.5
　115
　23
　34.5

❷ (1) 48.3　　(5) 129.6
(2) 67.2　　(6) 187.2
(3) 78.3　　(7) 146.2
(4) 83.2　　(8) 166.5

❸ (1) 216.6　　(4) 507.4
(2) 299　　(5) 582.8
(3) 405.6

■ Review of Step 27
(1) 44.2　　(3) 111.6
(2) 44.8

❶ (1) 　1.4
　　　× 0.2
　　　0.2 8

(2) 0.56
(3) 0.36
(4) 0.32
(5) 0.72

❷ (1) 0.92　(3) 0.78　(5) 0.72
(2) 0.96　(4) 0.68　(6) 0.84

❸ (1) 　6.4
　　　× 0.2
　　　1.2 8

(3) 　6.5
　　　× 0.4
　　　2.6 0

(5) 2.19
(6) 4.41
(7) 6.72
(8) 7.74

(2) 1.89　(4) 2.96

■ Review of Step 28
(1) 0.54　　(2) 4.48　　(3) 4.32

❶ (1) 　2.3
　　　× 2.6
　　　1 3 8
　　　4 6
　　　5.9 8

(3) 　　2.5
　　　×4.5
　　　1 2 5
　　　1 0 0
　　1 1.2 5

(2) 　2.4
　　×3.8
　　1 9 2
　　7 2
　　9.1 2

❷ (1) 20.88　(5) 28.14　(9) 46.72
(2) 24.91　(6) 29.44　(10) 56.44
(3) 32.66　(7) 26.86　(11) 56.28
(4) 28.22　(8) 32.16

■ Review of Step 29
(1) 17.92　　(3) 48.64
(2) 30.72

❶ (1) 　1.3 4
　　　×　2.8
　　1 0 7 2
　　2 6 8
　　3.7 5 2

(2) 　1.6 8
　　　×　2.6
　　1 0 0 8
　　3 3 6
　　4.3 6 8

❷ (1) 　3.4 6
　　　×　4.8
　　2 7 6 8
　　1 3 8 4
　1 6.6 0 8

(6) 　6.7 2
　　　×　6.2
　　1 3 4 4
　　4 0 3 2
　4 1.6 6 4

(2) 　3.4 6
　　　×　5.2
　　　6 9 2
　　1 7 3 0
　1 7.9 9 2

(7) 　6.8 4
　　　×　7.2
　　1 3 6 8
　　4 7 8 8
　4 9.2 4 8

(3) 　4.7 8
　　　×　5.6
　　2 8 6 8
　　2 3 9 0
　2 6.7 6 8

(8) 　7.6 8
　　　×　6.8
　　6 1 4 4
　　4 6 0 8
　5 2.2 2 4

(4) 　5.6 4
　　　×　5.4
　　2 2 5 6
　　2 8 2 0
　3 0.4 5 6

(9) 　8.4 6
　　　×　6.8
　　6 7 6 8
　　5 0 7 6
　5 7.5 2 8

(5) 　5.8 2
　　　×　6.4
　　2 3 2 8
　　3 4 9 2
　3 7.2 4 8

(10) 　9.5 8
　　　×　6.7
　　6 7 0 6
　　5 7 4 8
　6 4.1 8 6

Review of Step 20,21
(1) 0.8　　　　(3) 12
(2) 20.4　　　(4) 1700

Review of Step 22-25
(1) 34.2　　　(4) 88.2
(2) 105.6　　(5) 109.98
(3) 19.1　　　(6) 11.52

Review of Step 26
(1) 17.5　　　(2) 33.6

Review of Step 27
(1) 41.4　　　(2) 145.6

Review of Step 28,29
(1) 0.42　　　(4) 11.52
(2) 1.44　　　(5) 23.52
(3) 5.04　　　(6) 65.52

Review of Step 30
(1)
```
   2.86
×  2.8
  2288
  572
  8.008
```
(2)
```
   5.82
×  6.2
  1164
  3492
  36.084
```

■ Review of Step 30
(1) 14.532　　(2) 43.392

❶ (1) 1.2　　　(3) 4.2
(2) 2.1　　　(4) 3.2

❷ (1) 5.2　　(5) 3.1　　(9) 5.2
(2) 3.2　　(6) 4.2　　(10) 6.2
(3) 2.1　　(7) 6.3
(4) 4.2　　(8) 5.3

■ Review of Step 31
(1) 3.1　　　(2) 6.2

❶ (1) 2.4　　　(4) 1.3
(2) 1.2　　　(5) 1.2
(3) 1.7

❷ (1) 3.7　　(5) 3.2　　(9) 14.3
(2) 5.5　　(6) 4.2　　(10) 12.2
(3) 3.2　　(7) 17.8　　(11) 12.3
(4) 3.2　　(8) 14.9　　(12) 19.7

■ Review of Step 32
(1) 1.9　　　(2) 9.4

❶ (1) 0.2　　　(4) 0.2
(2) 0.1　　　(5) 0.3
(3) 0.3

❷ (1) 0.3　　(5) 0.6　　(9) 0.9
(2) 0.2　　(6) 0.2　　(10) 0.9
(3) 0.2　　(7) 0.4　　(11) 0.9
(4) 0.2　　(8) 0.5　　(12) 0.9

■ Review of Step 33
(1) 0.9　　　(2) 0.7

❶ (1) 2.36　　(2) 2.49

❷ (1)
```
    0.09
  2)0.18
    18
    0
```
(2) 0.09
(3) 0.08
(4) 0.07

❸ (1)
```
      2.35
   4)9.4
     8
     14
     12
      20
      20
       0
```

(3) 1.35

(4) 0.126

(2)
```
      0.92
   5)4.6
     45
      10
      10
       0
```

■ Review of Step 34

(1) 0.06   (2) 0.09

❶ (1)
```
       3.4
   18)61.2
      54
       72
       72
        0
```

(3)
```
      0.86
   15)12.9
      120
       90
       90
        0
```

(2)
```
      2.3
   23)52.9
      46
       69
       69
        0
```

(4)
```
      0.75
   18)13.5
      126
        90
        90
         0
```

❷ (1)
```
      1.42
   15)21.3
      15
       63
       60
        30
        30
         0
```

(5)
```
      2.35
   32)75.2
      64
       112
        96
        160
        160
          0
```

(2)
```
      1.08
   15)16.2
      15
       120
       120
         0
```

(6)
```
      3.15
   26)81.9
      78
       39
       26
        130
        130
          0
```

(3)
```
      2.45
   18)44.1
      36
       81
       72
        90
        90
         0
```

(7)
```
      1.284
   25)32.1
      25
       71
       50
        210
        200
         100
         100
           0
```

(4)
```
      2.24
   15)33.6
      30
       36
       30
        60
        60
         0
```

(8)
```
      0.675
   16)10.8
      96
       120
       112
        80
        80
         0
```

STEP 36 (P.82 · 83)

■ Review of Step 35
(1) 1.05   (2) 0.85

❶ (1)
```
        2
      0.17
   7)1.2
      7
      50
      49
       1
```
(2)
```
        4.4
   8)35.4
      32
       34
       32
        2
```

❷ (1) 0.6   (5) 0.1
(2) 0.6   (6) 0.2
(3) 2   (7) 0.2
(4) 2   (8) 0.05

STEP 37 (P.84 · 85)

■ Review of Step 36
(1) 0.3   (2) 0.3

❶ (1)
```
     0.133
   9)1.2
     9
     30
     27
      30
      27
       3
```
(2)
```
        3
     0.228
   7)1.6
     14
      20
      14
       60
       56
        4
```

❷ (1) 0.71   (5) 1.1
(2) 0.69   (6) 1.1
(3) 2.1   (7) 0.21
(4) 1.2   (8) 0.44

STEP 38 (P.86 · 87)

■ Review of Step 37
(1) 1.2   (2) 0.21

❶ (1)
```
      2
   3)7.2
      6
      1.2
```
(2) 2 R 4.7
(3) 3 R 1.3
(4) 6 R 2.5

❷ (1) 4 R 0.4   (6) 12 R 0.1
(2) 8 R 0.3   (7) 3 R 9.2
(3) 13 R 6.4   (8) 2 R 22.8
(4) 8 R 6.5   (9) 2 R 4.3
(5) 22 R 0.8   (10) 2 R 3.8

STEP 39 (P.88 · 89)

■ Review of Step 38
(1) 4 R 2.1   (2) 4 R 3.6

❶ (1) 2.6   (4) 3.4
(2) 2.8   (5) 3.6
(3) 3.2

❷ (1) 1.5   (7) 4.5
(2) 3.5   (8) 4.8
(3) 2.5   (9) 9.5
(4) 2.5   (10) 6.5
(5) 2.5   (11) 9.6
(6) 5.5   (12) 7.5

STEP 40 (P.90 · 91)

■ Review of Step 39
(1) 4.6   (2) 9.5

❶ (1) 0.5   (3) 0.5
(2) 0.6

❷ (1)
```
      0.8
  15)12
     120
       0
```
(5)
```
      2.5
  16)40
     32
     80
     80
      0
```

(2)
```
      0.5
  26)13
     130
       0
```
(6)
```
      0.48
  25)12
     100
     200
     200
       0
```

(3)
```
      0.5
  38)19
     190
       0
```
(7)
```
      2.08
  25)52
     50
     200
     200
       0
```

(4)
```
      0.6
  35)21
     210
       0
```
(8)
```
      1.25
  48)60
     48
     120
      96
     240
     240
       0
```

TEST
(P.92 · 93)

Review of Step 31
(1) 5.1   (2) 6.4

Review of Step 32,33
(1) 2.3   (2) 0.8

Review of Step 34
(1) 2.14   (2) 0.124

Review of Step 35
(1) 0.75   (2) 2.35

Review of Step 36
(1) 0.7   (2) 0.1

Review of Step 37
(1) 2.1   (2) 1.1

Review of Step 38
(1) 15 R 1.7   (2) 3 R 9.1

Review of Step 39,40
(1) 1.25   (2) 3.5

STEP 41
(P.94 · 95)

■ Review of Step 40
(1) 1.2   (2) 0.75

❶
(1) 20   (3) 20
(2) 30   (4) 20

❷
(1) 12   (6) 80
(2) 35   (7) 150
(3) 28   (8) 50
(4) 8   (9) 50
(5) 25   (10) 50

STEP 42
(P.96 · 97)

■ Review of Step 41
(1) 20   (2) 60

❶ (1) 2   (2) 3

❷
(1) 1.2   (6) 8
(2) 1.5   (7) 2.5
(3) 2.5   (8) 8.5
(4) 3.4   (9) 0.6
(5) 3.2   (10) 0.5

■ Review of Step 42
(1) 3　　　　　　(2) 3.5

❶ (1)
```
        1.7
  1,6)2,7.2
      1 6
      1 1 2
      1 1 2
            0
```

(2)
```
        3.4
  2,1)7,1.4
      6 3
        8 4
        8 4
            0
```

❷ (1)
```
        2.3
  2,5)5,7.5
      5 0
        7 5
        7 5
            0
```

(5)
```
        1.25
  1,8)2,2.5
      1 8
        4 5
        3 6
          9 0
          9 0
              0
```

(2)
```
        0.8
  2,7)2,1.6
      2 1 6
            0
```

(6)
```
        2.28
  3,5)7,9.8
      7 0
        9 8
        7 0
          2 8 0
          2 8 0
              0
```

(3)
```
        4.3
  3,2)13,7.6
      1 2 8
          9 6
          9 6
            0
```

(7)
```
        0.54
  2,5)1,3.5
      1 2 5
        1 0 0
        1 0 0
            0
```

(4)
```
        0.65
  5,6)3,6.4
      3 3 6
        2 8 0
        2 8 0
            0
```

(8)
```
        1.45
  3,2)4,6.4
      3 2
      1 4 4
      1 2 8
        1 6 0
        1 6 0
            0
```

■ Review of Step 43
(1) 2.3　　　　　(2) 0.52

❶ (1)
```
        3
  0,7)2,1
      2 1
        0
```

(3)
```
        5
  0,24)1,20
       1 2 0
            0
```

(2)
```
        1.5
  0,8)1,2
      8
      4 0
      4 0
        0
```

(4)
```
        5
  0,36)1,80
       1 8 0
            0
```

❷ (1)
```
        3
  0,03)0,09
       9
       0
```

(5)
```
        0.95
  0,8)0,7.6
      7 2
        4 0
        4 0
          0
```

(2)
```
        8
  0,05)0,40
       4 0
         0
```

(6)
```
        2.74
  0,5)1,3.7
      1 0
        3 7
        3 5
          2 0
          2 0
            0
```

(3)
```
        1.2
  0,75)0,90
       7 5
        1 5 0
        1 5 0
            0
```

(7)
```
        3.6
  0,65)2,34
       1 9 5
         3 9 0
         3 9 0
             0
```

(4)
```
        40
  0,31)12,40
       1 2 4
             0
```

(8)
```
        4.5
  0,86)3,87
       3 4 4
         4 3 0
         4 3 0
             0
```

■ Review of Step 44

(1) 5　　　　(2) 3.5

❶

(1)
```
          1.2
0.74)0.88.8
     7 4
     1 4 8
     1 4 8
         0
```

(2)
```
          2.5
0.35)0.87.5
     7 0
     1 7 5
     1 7 5
         0
```

❷

(1)
```
         2.1
1.24)2.60.4
     248
     1 2 4
     1 2 4
         0
```

(5)
```
        2.45
3.24)7.93.8
     648
     1458
     1296
     1620
     1620
        0
```

(2)
```
         1.3
1.45)1.88.5
     145
     435
     435
       0
```

(6)
```
        1.25
4.38)5.47.5
     438
     1095
     876
     2190
     2190
        0
```

(3)
```
         1.4
2.24)3.13.6
     224
     896
     896
       0
```

(7)
```
        1.24
4.25)5.27
     425
     1020
     850
     1700
     1700
        0
```

(4)
```
         2.3
3.14)7.22.2
     628
     942
     942
       0
```

(8)
```
        2.12
5.65)11.97.8
     1130
     678
     565
     1130
     1130
        0
```

■ Review of Step 45

(1) 3.1　　　　(2) 4.2

❶

(1)
```
        1.59
4.7)7.5
    47
    280
    235
    450
    423
     27
```

(2)
```
       3.34
2.6)8.7
    78
    90
    78
    120
    104
     16
```

❷

(1)
```
        3.46
2.6)9.0
    78
    120
    104
    160
    156
      4
```

(5)
```
       2.86
2.9)8.3
    58
    250
    232
    180
    174
      6
```

(2)
```
       83.33
0.9)75.0
    72
    30
    27
    30
    27
    30
    27
     3
```

(6)
```
        3.17
6.7)21.3
    201
    120
    67
    530
    469
     61
```

(3)
```
       2.60
9.6)25.0
    192
    580
    576
     40
```

(7)
```
       2.71
8.9)24.2
    178
    640
    623
    170
    89
    81
```

(4)
```
        9.08
3.7)33.6
    333
    300
    296
      4
```

(8)
```
        1.84
0.26)0.48
     26
     220
     208
     120
     104
      16
```

■ Review of Step 46
　（1）3.3　　　　　　　（2）1.4

❶
```
          2.430
  0.36)0.87.5
        72
        155
        144
        110
        108
         20
```

❷ （1）
```
          0.544
  1.34)0.73.0
        670
        600
        536
        640
        536
        104
```

（3）
```
          1.360
  6.57)8.94
        657
        2370
        1971
        3990
        3942
         480
```

（2）
```
            3
          1.825
  2.56)467.4
        256
        2114
        2048
         660
         512
        1480
        1280
         200
```

（4）
```
            8
          39.477
  0.09)355.3
        27
        85
        81
        43
        36
        70
        63
        70
        63
         7
```

■ Review of Step 47
　（1）3.17　　　　　　（2）1.89

❶ （1）
```
        7.3
  1.3)9.6
      91
      50
      39
      0.11
  (R 0.11)
```

（2）
```
        3.4
  3.2)10.9
      96
      130
      128
      0.02
  (R 0.02)
```

❷ （1）
```
        0.98
  0.9)0.8.9
      81
      80
      72
      0.008
  (R 0.008)
```

（5）
```
        2.1
  1.24)261.6
       248
       136
       124
       0.012
  (R 0.012)
```

（2）
```
        0.13
  6.5)0.8.9
      65
      240
      195
      0.045
  (R 0.045)
```

（6）
```
        2.3
  0.12)0.28.7
       24
       47
       36
       0.011
  (R 0.011)
```

（3）
```
        3.1
  7.2)22.7.4
      216
      114
      72
      0.42
  (R 0.42)
```

（7）
```
        2.4
  2.46)592.5
       492
       1005
       984
       0.021
  (R 0.021)
```

（4）
```
        4.2
  7.3)30.7.8
      292
      158
      146
      0.12
  (R 0.12)
```

（8）
```
        3.1
  3.82)1185.4
       1146
       394
       382
       0.012
  (R 0.012)
```

Review of Step 41
(1) 20   (2) 28

Review of Step 42
(1) 6   (2) 2.8

Review of Step 43
(1) 1.8   (2) 1.62

Review of Step 44
(1) 5   (2) 3.5

Review of Step 45
(1) 2.15   (2) 1.46

Review of Step 46,47
(1) 2.1   (2) 1.4

Review of Step 48
(1) 6.7 R 0.12
(2) 2.1 R 0.012